Global Issues Series

General Editor: **Jim Whitman**

This exciting new series encompasses tl
human and natural systems; cooperation a
The series as a whole places an emphasis on
causal relations in political decisionmaking; ...y, control and accountability in issues of scale; anc ... conflicting values
and competing claims. Throughout the series ...centration is on an integration
of existing disciplines towards the clarification of political possibility as well as
impending crises.

Titles include:

Roy Carr-Hill and John Lintott
CONSUMPTION, JOBS AND THE ENVIRONMENT
A Fourth Way?

Malcolm Dando
PREVENTING BIOLOGICAL WARFARE
The Failure of American Leadership

Brendan Gleeson and Nicholas Low (*editors*)
GOVERNING FOR THE ENVIRONMENT
Global Problems, Ethics and Democracy

Roger Jeffery and Bhaskar Vira (*editors*)
CONFLICT AND COOPERATION IN PARTICIPATORY NATURAL RESOURCE MANAGEMENT

Ho-Won Jeong (*editor*)
GLOBAL ENVIRONMENTAL POLICIES
Institutions and Procedures
APPROACHES TO PEACEBUILDING

W. Andy Knight
A CHANGING UNITED NATIONS
Multilateral Evolution and the Quest for Global Governance

W. Andy Knight (*editor*)
ADAPTING THE UNITED NATIONS TO A POSTMODERN ERA
Lessons Learned

Kelley Lee (*editor*)
HEALTH IMPACTS OF GLOBALIZATION
Towards Global Governance

Nicholas Low and Brendan Gleeson (*editors*)
MAKING URBAN TRANSPORT SUSTAINABLE

Graham S. Pearson
THE UNSCOM SAGA
Chemical and Biological Weapons Non-Proliferation

Andrew T. Price-Smith (*editor*)
PLAGUES AND POLITICS
Infectious Disease and International Policy

Michael Pugh (*editor*)
REGENERATION OF WAR-TORN SOCIETIES

Bhaskar Vira and Roger Jeffery (*editors*)
ANALYTICAL ISSUES IN PARTICIPATORY NATURAL RESOURCE MANAGEMENT

Simon M. Whitby
BIOLOGICAL WARFARE AGAINST CROPS

Global Issues Series
Series Standing Order ISBN 978-0-333-79483-8
(*outside North America only*)

You can receive future titles in this series as they are published by placing a standing order. Please contact your bookseller or, in case of difficulty, write to us at the address below with your name and address, the title of the series and the ISBN quoted above.

Customer Services Department, Macmillan Distribution Ltd, Houndmills, Basingstoke, Hampshire RG21 6XS, England

---

Making Urban Transport Sustainable

Edited by

Nicholas Low
Associate Professor in Environmental Planning
Faculty of Architecture, Building and Planning
University of Melbourne
Australia

and

Brendan Gleeson
Deputy Director, Urban Frontiers Program
University of Western Sydney
Australia

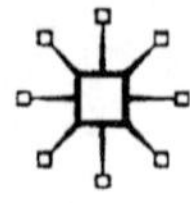

Softcover reprint of the hardcover 1st edition 2003 978-0-333-98198-6

First published 2003 by
PALGRAVE MACMILLAN
Houndmills, Basingstoke, Hampshire RG21 6XS and
175 Fifth Avenue, New York, N.Y. 10010
Companies and representatives throughout the world

PALGRAVE MACMILLAN is the global academic imprint of the Palgrave Macmillan division of St. Martin's Press, LLC and of Palgrave Macmillan Ltd. Macmillan® is a registered trademark in the United States, United Kingdom and other countries. Palgrave is a registered trademark in the European Union and other countries.

ISBN 978-1-349-43035-2 ISBN 978-0-230-52383-8 (eBook)
DOI 10.1057/9780230523838

This book is printed on paper suitable for recycling and made from fully managed and sustained forest sources.

A catalogue record for this book is available from the British Library.

Library of Congress Cataloging-in-Publication Data
Making urban transport sustainable/edited by Nicholas Low and Brendan Gleeson.
p. cm. – (Global issues series)
Includes bibliographical references and index.
1. Urban transportation – Environmental aspects – Case studies. 2. Sustainable development – Case studies. I. Low, Nicholas. II. Gleeson, Brendan, 1964– III. Global issues series (Palgrave Macmillan (Firm)) P302.C6858 2002

HE305 .M35 2002
388.4'042–dc21 2002028747

This book is dedicated to people who go on foot

In hardly any city in the world is the urban transportation system optimal.

(Hall and Pfeiffer, 2000: 21)

The crucial issue is ... not a quest for 'sustainable transport' as some independently defined goal. Rather, the relevant ultimate target is overall sustainability, implying certain necessary features of transport.

(Button and Verhoef, 1998: 335)

[These quotations were provided by Rolf Lidskog, Ingemar Elander and Pia Brundin.]

Contents

List of Figures, Tables and Boxes

Notes on the Contributors

Swapna Banerjee-Guha is Professor of Human Geography at the University of Mumbai, India. Dr Banerjee-Guha is a member of the panel of doctoral examiners of the University of London and a Steering Committee member of the International Critical Geography Group. She has taught and participated in seminars in the USA, the UK and various European countries. She was visiting fellow at the Institute of Development Economics in 2001. Her research areas include urban, social and development issues at national and international levels.

Pia Brundin has a Master's degree in political science and is a PhD candidate in political science at Örebro University. She has participated in research projects on environmental politics and on the implementation of Agenda 21 in Swedish municipalities. She is currently working on a dissertation on how environmental and other transnational social movements use the Internet to further their politcal goals at a global level.

Paul Barter has a PhD from Murdoch University in the comparative urban transport field, focusing on the problems of high density Asian density Asian cities. Paul ran the Sustainable Transport Action Network for Asia and the Pacific (the SUSTRAN Network) for a number of years. He collected data for Asian cities for the UITP Millennium Cities Database and he is now working as a Lecturer in the Geography Department of the National University in Singapore.

C.J. Campbell has a PhD in geology from Oxford University and joined the oil industry in 1958. His early years were spent as a field geologist and, in 1968, he joined the head office of an oil company in New York, becoming regional geologist for South America. After a two-year assignment as Chief Geologist in Ecuador, he went to Norway for ten years as exploration manager and later executive vice-president of major oil companies. Since 1990 he has written two books and numerous articles on the study of oil depletion, as well as lecturing, broadcasting, and assisting in the production of TV programmes. Finally, he was instrumental in the formation of ASPO (The Association for the Study of Peak Oil), which is a network of concerned scientists representing most European countries, now influencing governments and the European Union.

Carey Curtis is a lecturer and researcher at Curtin University, Western Australia, Carey Curtis is also consultant and adviser to government. She has worked on a wide range of projects in the planning and transport fields for local, state and central governments in both Australia and the UK. Her research interests have focused on sustainable transport and include land use and transport integration through sustainable urban design, accessibility planning, travel behaviour, and travel demand management. She is also involved in sustainable transport through local advocacy groups.

Carlos Destefani is a Research Engineer at the Industrial Research Institute Swinburne (IRIS), Swinburne University of Technology, Melbourne, Australia.

Ingemar Elander is Associate Professor in Political Science and Director of the Centre for Housing and Urban Research, Örebro University, Sweden. He is author and co-author of several articles and book chapters on urban governance, social democracy and planning, environmental and housing policy, and related themes. His latest article to be published is 'Partnerships and urban governance' for the *International Social Science Journal*, 172, 2002.

Gang Hu is a research scholar at The University of Melbourne. Gang Hu has published widely in China and is co-author of five books on urban transport, planning and urbanization. He is the winner of a Science and Technology Progress Award from the Government of Zhejiang for his work on the planning of Hangzhou.

Brendan Gleeson is Deputy Director of the Urban Frontiers Programme, University of Western Sydney. His research and publication is in the field of political economy of planning, urban social policy, and environmental theory and policy. His book *Geographies of Disability* was published in 1998. He has worked with Nicholas Low on a number of projects. Their book *Justice, Society and Nature* (1998) won the 1998 Harold and Margaret Sprout award for political ecology. *Australian Urban Planning* (2000) won the Planning Institute of Australia national award for scholarship.

Jeff Kenworthy is Associate Professor in Sustainable Settlements in the Institute for Sustainability and Technology Policy at Murdoch University in Perth, Western Australia. He has spent 23 years in the urban planning and transport field. His main contributions have been extensive publications on automobile dependence in cities and how to reduce it, with an emphasis on international comparisons between cities. He is co-author of the *Millennium Cities Database for Sustainable Transport,* which compares 100 cities worldwide.

Felix Laube has a PhD from Murdoch University in the comparative urban transport field and has worked for 3 years as a Senior Researcher in the

Institute for Sustainability and Technology Policy, Murdoch University on the UITP *Millennium Cities Database for Sustainable Transport.* He is currently employed by Swiss Federal Railways.

Rolf Lidskog is Professor in Sociology at the Centre for Housing and Urban Research, Örebro University. He conducts research on environmental policy and politics at international and national level, especially the role of expertise in environmental politics. Among his recent contributions is the co-edited book *Consuming Cities: the Urban Environment in the Global Economy after the Rio Declaration* (eds N. Low, B. Gleeson, I. Elander and R. Lidskog, 2001).

Nicholas Low is Associate Professor in Environmental Planning at the Faculty of Architecture, Building and Planning, The University of Melbourne, Australia. He is the author or editor of seven books, Fellow of the Royal Australian Planning Institute, a member of the editorial team of the international journal *Urban Policy and Research* and a consultant to the Volvo Educational and Research Foundation.

Fujio Mizuoka took his doctorate at Clark University Massachusetts USA where he was a Fulbright Scholar. He was a visiting lecturer at the University of Hong Kong and from 1992 has been Professor of Economic Geography at Hitotsubashi University, Tokyo, Japan.

Peter Newman is Professor of City Policy and Director of the Institute for Sustainability and Technology Policy, Murdoch University, Perth, Australia. Visiting Professor at the University of Pennsylvania. Professor Newman is a consultant to many governments internationally and the State of Western Australia.

Elias Siores is Professor of Mechanical Engineering and Executive Director of the Industrial Research Institute Swinburne, IRIS, Swinburne University of Technology, Melbourne, Australia. Honourary Professor, Tsinghua University, Beijing, People's Republic of China.

Hartmut Stiller is a physicist but also trained in economics. From 1992 until 2001 Hartmut Stiller was senior fellow at the Wuppertal Institute for Climate, Environment and Energy in the Department of Material Flows and Structural Change. His main fields of research are in measuring material flows and resource productivity in the fields of transport and new technologies; globalization and the environment, in particular in the area of trade and environment.

Izumi Takeda holds a Master of Education degree. He is a lecturer at Hokkaido University of Education at Iwamizawa. His research interests include critical appraisal of transport policies in Japan, particularly after its privatization in 1987.

Emin Tengström received a PhD in Latin in 1964; was Assistant Professor in Latin at Göteborg University, Sweden in 1971; Chairman of the board of 'The Centre for Interdisciplinary Studies of the Human Condition' in 1972; Professor of Human Ecology at the Technical University of Chalmers, Göteborg in 1981; Professor of Human Ecology at Göteborg University in 1985; and Research Professor of Transportation Studies at Aalborg University, Denmark 1995.

John Whitelegg is Professor of Environmental Studies at Liverpool John Moore's University, UK and Visiting Professor of Transport at Roskilde University, Denmark. Professor Whitelegg is founder and managing director of Eco-Logica, a Lancaster based consultancy specializing in sustainable transport, and editor of the international journal *World Transport Policy and Practice*.

Cameron Yee is Policy Director at People United for a Better Oakland (PUEBLO). Before working at PUEBLO, he was the Transportation Director at the Urban Habitat Program writing reports on topics from transportation justice to gentrification. He was also a postgraduate researcher at the Institute of Transportation Studies at the University of California at Davis.

Internet Websites

(**Bold** characters denotes word for alphabetical listing)

Aalborg Charter of European Cities and Towns Towards Sustainability: http://www.iclei.org/europe/echarter.thm

Axcess **Australia**: http://www.axcessaustralia.com/car/power.asp

BATLUC (Bay Area Transportation and Land Use Commission): http://www.transcoalition.org/

Beijing Environmental Protection Bureau: http://www.bjepb.gov.cn/English_homepage/index.htm

BP: http://www.bp.com/environ_social/environment/ climate_change/ index.asp

Canadian Climate Change Calculator for personal greenhouse gas emissions: http://www.climcalc.net

Carfree Cities: http://www.carfree.com/

Centre for Alternative Technology, UK http://www.foe.co.uk/CAT

Cities for Climate Protection http://www.iclei.org/co2/

Cities As Sustainable Ecosystems (UNEP)

http://www.communityzero.com/unep-case/index.cfm?key = 654-NYL

CIVITAS: http://civitas.barcelona2004.org/

Critical Mass: http://www.criticalmasshub.com/

US **Fuel** Cell Council: http://www.usfcc.com/internet.htm

Herman **Daly**: http://iee.umces.edu/miiee/HERMANCV.html

Debunking **Economics**: http://www.debunking-economics.com

ELTIS (European Local Transport Information Service): www.eltis.org

US **Environmental** Protection Agency: EPAhttp://www.epa.gov/

European Commission: http://europa.eu.int/comm/index_en.htm

European Commission White Paper on Transport: http://europa.eu.int/comm/energytransport/en/lb_en.html

European Conference of Ministers of Transport (sustainable mobility in cities): www.inforegio.cec.eu.int

European Environment Agency: http://www.eea.eu.int/

Trans-**European** Networks: http://europa.eu.int/pol/ten/index_en.htm

Friends of the Earth: www.foe.co.uk

Gothenburg European Council: http://europa.eu.int/comm/gothenburg_council/index_en.htm

Hounslow, London Borough (air pollution monitoring): (www.hounslow.gov.uk/es/monitor.html)

Intergovernmental Panel on Climate Change: http://www.ipcc.ch/

International Council for Local Environmental Initiatives (ICLEI) http://www.iclei.org/

International Energy Agency: http://www.iea.org/

Japan Electric Vehicle Association (JEVA); http://jeva.or.jp
Japan Railways Group: http://www.japanrail.com/
The **Hadley** centre for Climate Prediction and Research: http://www.met-office.gov.uk/research/hadleycentre/
Ford **Hybrid** Vehicle, http://www.hybridford.com/index.asp,
Steve **Keen**: 'Debunking Economics': http://bus.uws.edu.au/Steve-Keen/
http://europa.eu.int/comm/gothenburg_council/index_en.htm
Maharashtra (India) Maps: http://www.mapsofindia.com/maps/maharashtra/
OECD: http://www.oecd.org/
Peak of **Oil**: http://www.hubbertpeak.com/campbell/
Toyota **Prius** http://prius.toyota.com
Rocky Mountains Institute (Amory and Hunter Lovins): http:// www.rmi. org/
Smart Growth America: http://smartgrowthamerica.org/
SUSTRANS (http://www.sustrans.org.uk/webcode/home.asp
TEA 21 (Transportation Equity Act for the 21st Century): http://www.fhwa.dot.gov/tea21/legis.htm
Trans-European Networks (TEN): http://europa.eu.int/pol/ten/ index_en.htm
Toyota (electric vehicles): http://www.toyota.com
UITP Database: http://wwwistp.murdoch.edu.au/publications/projects/uitpmill/uitpmill.html
UNCHS (UN Conference on Human Settlements, Habitat): http://www.unhabitat.org/
US Environmental Protection Agency: EPAhttp://www.epa.gov/
US Fuel Cell Council: http://www.usfcc.com/internet.htm
Wuppertal Institute for Climate, Environment and Energy: http://www.wupperinst.org/
Zero and Low Emissions in Urban Society (**ZEUS**): http://www.zeus-europe.org/

Acknowledgements

This project was conceived following a sabbatical visit in 1999 by Nicholas Low to the University of Amsterdam. This book is the outcome of international research by many authors and their colleagues. The editors wish to thank the many people, too numerous to mention, who have contributed their time and expertise to make the project a success. The editors of this volume wish to acknowledge particularly the support of the Urban Planning Program of the Faculty of Architecture, Building and Planning of the University of Melbourne, and the Urban Frontiers Program of the University of Western Sydney. Both universities provided Research Development Grants which assisted with the research on the Australia chapter especially. In this respect, the authors of Chapter 12 wish to acknowledge the work of Ms Emma Rush who worked as research assistant on the project. The Faculty of Architecture, Building and Planning (University of Melbourne) provided financial support for the making of the index to the volume. Ms Chandra Jayasuriya redrew the map in Figure 9.1b and drew the maps in Chapter 12.

Dr Emin Tengström wishes to acknowledge permission from Ashgate Ltd, the publishers of his book *Towards Environmental Sustainability: A Comparative Study of Danish, Dutch and Swedish transport policies in a European context*, for the use in Chapter 8 of this volume of tables 6.3 (p.175), 6.8 (p.178), 6.10 (p.180), 6.13 (p.182) and 6.14a (p.184) from the book.

The authors of Chapter 15 wish to acknowledge the financial support of the International Union (Association) of Public Transport (UITP) in Brussels, in developing the data used in this book. Mr Jean Vivier of UITP is especially thanked for his collaboration with the authors in carefully checking and verifying data in each city. We also wish to acknowledge the help of Michelle Zeibots, Gang Hu, Momoko Kumagai, Chamlong Poboon, Benedicto Guia (Jr) and Antonio Balaguer in helping to collect data in specific cities. SYSTRA in Paris also provided some initial starting data on 40 cities in the study. Finally, but not least, we wish to thank the hundreds of people worldwide who cooperated with the authors by providing data and assisting with innumerable requests for clarification and follow-up, to ensure all data were of the best available quality.

1

Is Urban Transport Sustainable?

Nicholas Low

Introduction

Once the privilege of the elite, personal mobility is a freedom bestowed by modernity on the general public through technology. The physical negotiation of space by people in pursuit of social values (access to work, friends, child care, education, recreation and supplying the home) is part of urban social life. Road freight vehicles provide a flexibility of supply that keeps profits up and costs down for businesses in the 'consuming' city. But the benefits of freedom and flexibility are illusory if the opportunity costs of providing for unending mobility are never considered, distances to be covered increase, travel becomes a compulsory, stressful, dangerous and expensive routine and the costs of mobility are merely shifted from the individual to society and the environment. This book explores how the real benefits of mobility can be protected and the costs properly allocated and contained. The chapters examine the sustainability of the world's urban transport systems, bringing a variety of perspectives from different nations and from different fields: engineering, sociology, critical geography, environmental economics, eco-politics, urban planning and transport planning.

What does 'sustainability' mean? The question will be approached from a familiar perspective, that of the triad – economic, social and environmental sustainability. In the 'triple bottom line' variant, this perspective requires that corporations and governments seek the simultaneous achievement of three fundamental goals: economic profitability, social responsibility and environmental conservation (Elkington, 1998). Our analysis shows, however, that such an outcome is contingent on power and commitment. There is no necessary correspondence between economic, social and environmental sustainability. Useful though the triple bottom line framework is as an accounting tool, sustainability will require a massive and concerted effort of political will and technical ingenuity, and a true 'paradigm shift' in the belief systems and education of engineers, urban planners and economists – the professional shapers of the city.

That shift is already under way. Many of the new generation of engineers, planners and economists are developing a new way of thinking about the infrastructure of the city, and the connection between transport, land use, society and environment. The new paradigm is developing 'critical mass' (to use Whitelegg's expression), enough momentum to make a real difference to transport planning policy and practice. In this book we hope to help the shift along a little further.

The paradox of sustainability

As every engineer knows there is nothing more stable than a triangle in which the forces conducted through each member cancel each other out (Figure 1.1).

Unfortunately applied to 'sustainability' this figure is paradoxical because for the environment to be sustained, both society and economy have to *change*. They cannot therefore be *sustained* in their present form. Debates about sustainability are fuelled by the perception that 'we can't go on as we are' (Blowers, 1995), that 'business as usual' can't continue (Athanasiou, 1998; Sachs, 1999), that society must find ways of curbing consumption even while spreading the capacity to consume more widely throughout the world (Daly, 1996). Peters (2000: 113) in a discussion of Europe's contribution to the 2002 Johannesburg Earth Summit writes: 'Accessibility and mobility gains are often reaped at the expense of severe damage to human health and global biodiversity, with problems accelerating for future generations. This clearly violates the sustainability principle'. White (2002: 57) citing research by Boardman *et al.* (1997) notes that by 2010, in the UK, carbon dioxide emissions from road transport are expected to rise to 27 per cent of total emissions, rivalling that of all industry. Urban transport is a key element in the structure of the city, and the city – and city network – is typically the geographical and social formation in which most people in the world are coming to relate to and consume the natural environment (Low *et al.*, 2000).

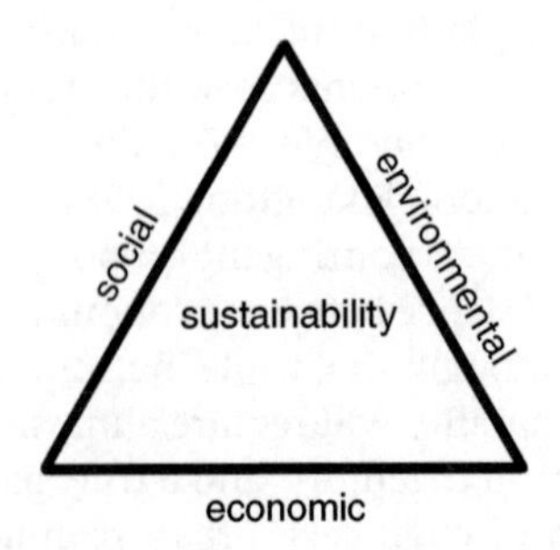

Figure 1.1 Sustainability as a triangle of forces. (Figure drawn by Chandra Jayasuriya.)

As this book will show, much urban transport policy today defies sustainability. For the environment to be sustained, urban transport has to change.

But let us probe this paradox a little more deeply. It cannot mean that the natural environment is unchanging or ever should be. But normal evolutionary environmental change must be distinguished from catastrophic change. Evolutionary change moves extremely slowly by the measure of a human lifespan. The rate of species extinction and global warming over the last two or three hundred years is comparable with what are regarded by palaeontologists as catastrophic events in the geological record. The contemporary event, however, is being driven by human exploitation. It is because of our growing understanding of catastrophic environmental change that changes in society are necessary to slow it down. How far these changes must go is of course a matter of debate.

The global biosphere is a closed system with one absolutely predominant energy input: solar radiation, and no output for waste. This system is composed of a variety of natural ecosystems of different scales (Applin *et al.*, 1999). Human beings cannot act significantly, individually and directly on the natural environment viewed as a system of ecosystems. The scale difference between a human individual and an ecosystem is just too big. Of course an individual may chop down a tree or recycle a can but, on its own, this act has little effect on the natural environment. Humans act collectively through society and its institutions, the most powerful of which are states and market economies (Lindblom, 1977; March and Olsen, 1989). These institutions organize and combine the many individual acts into social activity of such a scale that it leads to catastrophic deterioration of the environment.

The individual car driver is correct to conclude that one trip, even that individual's annual summation of trips in the car, will have an insignificant effect on the environment. Yet this one trip could not take place at all without society producing the wherewithal to make it: through mines, factories, machines, cars, roads, prices, taxes, regulations. And if society makes it possible for one individual, it also makes it possible for millions of others. The transportation patterns caused by the collective activity of millions of individuals feed back into other patterns created socially, for example patterns of land use, methods of distribution of goods, patterns of distribution of social opportunities, health, disease and death. Some patterns, especially patterns of production of goods and services and distribution of land uses, feed back into the means of transport. Worldwide, billions of trips in vehicles burning fossil fuel, are changing the Earth's climate, and therefore the environment of all species, including the human species (Whitelegg, 1997; IPCC, 2001).

The key to understanding sustainability is therefore twofold: first that it is through social institutions and mechanisms that individuals have a significant effect on the environment. Second, that humans are capable of changing society and its institutions. If society, and its mechanisms such as urban transport, were not subject to human will in some sense, there would be little point in thinking about the future and sustainability. An orientation to the future and to deliberate social change is a central feature of modernity and the technical application of science. For this reason governments deliver 'policies' looking to the future on which they are publicly judged.

The paradox of sustainability emerges strongly in urban transport. Governments that espouse environmental sustainability also espouse transport policies that deplete and ruin the environment. All too often in policy documents where environmental sustainability is discussed the urgent necessity of social and economic change is glossed over in language designed to reassure the reader that the right *trade-off* (or balance of forces in the triangle of Figure 1.1) can be found between economic, social and environmental policies. With this magic mixture all will indeed be for the best in this, the best of all possible worlds. Such policies deserve the term 'greenwash' (see Athanasiou, 1998: 238). Whereas it is possible to trade off environmental and health *benefits* against economic growth (for that is what has always been done), it is logically impossible to trade-off environmentally *unsustainable* growth against environmental *sustainability*. Growth is either environmentally sustainable or it's not.

Nevertheless, provided that the need for *change* of social and economic institutions is retained, the idea of the environmental, social and economic *dimensions* of sustainability is an important and fertile one. The order of consideration is significant. The economy is essentially a creation of society and should be understood as framed by rather than, as is more often assumed, framing society. Both operate within the context of a natural environment of limited capacity that can be reduced but not enlarged.

Urban transport and economic sustainability

Sustaining the economy has been the principal task of governments throughout the developed world since the end of the Second World War. It's easy to see why investment in transport infrastructure is such an appealing way of spending taxes and borrowed funds. It addresses a very concrete problem experienced by voters: traffic congestion. It supports the growth of personal mobility. Investments in roads seem to realize, in an immediate sense, the potential of the private car which has acquired more than just practical value for its owner (Sheller and Urry, 2000). Investment in public transport can be argued to reduce road traffic congestion, as in the famous case of London's Victoria Underground Line. Now 'sustainability' itself is being brought in to plead the case for new transport infrastructure.

Investment in transport supports business in two ways. It acts as a stimulus to the large sectors of industry involved in construction and transport. And it offers to reduce the costs of all businesses in transporting goods.

It appeals to the orthodox economists who advise governments. It doesn't require any real change in government thinking and therefore appears to carry minimum short-term political risk. In short, investment in transport infrastructure serves the immediate interests and beliefs of voters, businesses, bureaucrats and politicians. The support of professional engineers, trades unions, and sectors of the investment industry add further weight to the nexus of arguments in favour of transport investment as a means of sustaining the economy.

In order to sustain the economy and enhance the productivity of private capital, it is argued, there is a need for investment in transport infrastructure (see, for example, Aschauer, 1989, 1990; EC, 1993). Of course such investment has a multiplier effect from the construction involved, but theoretically the causal link between transport infrastructure *per se* and economic growth remains to be proved (Gramlich, 1994; Vickerman, 1998: 132). Rothengatter (1994: 116) claims as an external benefit of transport infrastructure: 'improving accessibility to remote regions may foster regional development and produce interregional multiplier effects'. Vickerman (1998: 153) argues that the relationship between infrastructure and economic growth is a complex one: 'Often it is intra-regional variations and access to local and regional networks which may be as critical for locations in peripheral regions as their actual peripherality'. But, as Whitelegg points out in Chapter 7, such empirical evidence as there is suggests no geographic link between transport investment and economic success. Black (2001: 2) regards the notion as a myth. While there are situations in the developing world (remote regions) where transport infrastructure investments may lead to economic growth 'it is unlikely that these conditions exist any longer for [the developed world]'.

The proposition that arguments in favour of transport investment are specious and contradictory tends to be eclipsed by the fact that, well, the world economy *has* been sustained. The world has not experienced a Great Depression on the scale of that of the 1930s, a global economic crisis with vast social and political repercussions – not yet. In view of that fact, and in view of the supporting interests of the key players in the economy and politics, what is perhaps surprising is that transport policy *is* changing, however slowly (see Chapters 2, 6, 13 and 14). It is changing at least partly because there is a dawning awareness in some departments of government and business that something is wrong with the economic calculations. The 'something' is the assumption underlying those calculations that the environment is an unlimited cornucopia. It is as if orthodox economists have looked at the environment through the eyes of an individual and have seen that its scale is so vast that nothing the individual can do will have any appreciable impact on it. The sustainability of an economy without environmental limits is one thing, but what if the economy as a social whole becomes so big that it actually presses up against the limits of what the environment can provide?

Transportation economics is concerned mainly with efficiency and allocation (Button, 1977; Small, 1992). An efficient allocation is one in which the kind of commodities, and how much of each, is produced coincides with what and how much is demanded under any given distribution of the capacity to pay in the population. Normatively speaking, what type of and how much transportation is provided should be no more than and no different from what is demanded by the aggregate of consumers (under budget constraints) when each weighs up all the benefits and costs to him or her individually. The problem is how to get all the costs bearing down on the individual (or organisation) who receives the benefits, that is to internalize the costs.

From the point of view of environmental sustainability, this sounds rather good in principle because it minimizes the use of resources for a given social outcome. Road pricing to internalize the costs of congestion is the classic case and has been praised by environmentalists (Jacobs, 1991: 146). Arguing in favour of the use of economic instruments in infrastructure provision (as opposed to infrastructure use), Rothengatter (1994) posits that transport infrastructure is a 'club good' in that its supply stimulates agents to seek additional 'rents' by free riding and 'as a consequence, excess demands for infrastructure capacity are generated' (122). This merely recognizes theoretically what has already been established empirically, which in turn was suggested by common sense: that building better road systems generates additional demand to use them (see SACTRA, 1994).

However, there is a more compelling reason why economists applaud technical infrastructure solutions. The political value at the heart of microeconomics is the belief in the virtue of competition. Space limits competition and interferes with the perfection of the market. The larger the population that can be brought into competition, the better. The connection between transportation, the growth of business and the growth of cities was discussed in one of the seminal texts of urban geography *The City* by Park, Burgess and McKenzie. Transportation systems are seen naturalistically as a part of human ecologies. McKenzie quotes Hadley: 'It is this quickening and cheapening of transportation that has given such stimulus in the present day to the growth of large cities. It enables them to draw cheap food from a far larger territory and it causes business to locate where the widest business connection is to be had, rather than where the goods or raw materials are most easily produced' (McKenzie, 1967: 69 n2).[1]

Transport negates the effect of space and enlarges the competing population of firms and households by bringing more people and products within reach of one another. This view of the world infuses the publications of the European Commission (see Chapter 13 of this volume). The American economist Paul Krugman (1999: 85) writes with delight about 'globalization' which puts fresh vegetables from Zimbabwe on the tables of Londoners. Whitelegg writes *not* with delight but dismay about the 150 gram pot of strawberry yoghourt which is responsible for moving one lorry 9.2 metres in the process of production (Whitelegg, 1997: 39 citing research by Böge, 1995). Both are tales of spatial expansion and flexibilisation, principles of the globalization of the market economy: the ever increasing choice of location of production of anything and everything.

Economic rationality contains two kinds of problem for transport sustainability, one external, to do with the subject matter (transport), and one internal to the theory of economics itself. The external problems have to do with the nature of transport. First, the disaggregation entailed in the market mechanism tends to work against the technical co-ordination (intermodal, intertemporal and geographical) required by an effectively functioning transport system. Rothengatter (1994: 125) points to the problem of natural monopolies in transport networks, but the matter goes deeper than that. The technical efficiency of a transport system depends on the smooth integration of all its elements (Vuchic, 1999: 295). For instance, feeder bus timetables must be co-ordinated with train timetables. Ticketing systems must be network-wide. Encouragement of transit use must be accompanied by encouragement of walking and cycling, and discouragement of car use for journeys where transit supplies a viable alternative.

Equally importantly, while some parts of the use of transport systems can be subjected to market mechanisms, including the use of roads, the *creation* of transport systems cannot – or at least *is* not. Transport infrastructure investment is enormously lumpy, it has very long-term effects on the land use and production process, social needs do not quickly adapt to changes in transport provision and, while benefits are individualized, transport systems incur huge external environmental costs, some of which are almost unmeasurable (the cost of greenhouse gas emissions for example). If all the social and environmental costs of road systems were actually internalized in a price for road use, the result would probably be that only the rich could use the roads, thus violating social sustainability (see below). Second, increasing transport infrastructure feeds back into more spatially dispersed patterns of land use. Land use patterns have been shaped over decades by government provision of infrastructure to accommodate motor vehicles. If economizing on transport is needed – as it is – it must be led by government with infrastructure and logistical policies which enable such economizing.

The second kind of problem is internal. Again it is twofold. At microeconomic level, the theory does not account adequately for time, nor does it acknowledge the ultimate limits imposed by the capacity of the environment. In the real world economic processes take time yet, as Keen (2001: 166) points out, 'economists don't consider time in analysing demand, supply, or any of their other key variables'. The conventional economic approach to time goes no further than the rate at which individuals discount the future, and much space in environmental economics texts is taken up with mathematical representations of this quite marginal theoretical problem (see for example Pearson, 2000, chapter 4).

It doesn't take much thought to see that asking whether an actual person living now would rather have more mining and less coral reefs is not the most useful line of questioning from the viewpoint of global ecology and resource depletion. As Martinez-Alier (1987: 171) pointed out, the critique is that 'economic theory is unable to deal with inter-generational allocation of exhaustible resources relying only on the exchanges between agents whose behaviour accords with the postulate of rationality and utilitarian calculus because non-born agents cannot bid in today's market'. But then economists today do not usually ask people questions, they make assumptions and then carry out mathematical operations on the basis of those assumptions. One of the assumptions Pearson makes is that 'avoiding current costs and securing near-term benefits frees up resources in the near term that can be reinvested at a positive rate, thus increasing the wealth of future generations' (Pearson, 2000: 100). This ignores the distribution of benefits, and assumes both that human-made capital is a substitute for natural capital and that human benefit is the only benefit that matters ethically. For instance, no amount of human products like cars or trains can compensate for the death of a biological wonder such as the Great Barrier Reef. Of course this assertion may be disputed but the questions here are ethical not economic.

The discount rate merely considers time preference from a static point, but the core theoretical economic issue is dynamic. Keen (2001) uses a vivid image to illustrate the problem. Imagine you have never seen or heard of a bicycle and that a bike guru convinces you that there are two steps to learning to ride: 'in step 1, you master balancing on a stationary bike. In step 2, you master riding a moving bike, applying the skills acquired at step 1' (165). Learn like that and the moment you try and change direction you fall flat on the road. The forces at work, and the skills needed to master them are different in a dynamic state from a static one. Stringing together a series of static analyses does not amount to dynamic analysis. Keen explains the problem very clearly and exhaustively in chapter 8 of his book (pp. 165–87; see also: http://bus.uws.edu.au/Steve-Keen/). The failure to deal with time means that microeconomics tends to ignore real feedback effects such as the land use dispersing effect of infrastructure.

Conventional macroeconomics ignores the limited capacity of the Earth's environment. Overcoming space by technology is today stretching the capacity of the environment to absorb the wastes thereby produced. Daly explains in terms of a model of the economy in which raw material is taken in from the environment, put through the economic system, and transformed into non-material human satisfactions and material waste:

> Ecological sustainability of the throughput is not guaranteed by market forces. The market cannot by itself register the cost of its own increasing

> scale relative to the ecosystem. Market prices measure the scarcity of individual resources relative to each other. Prices do not measure the absolute scarcity of resources in general.... Ecological criteria of sustainability, like ethical criteria of justice, are not served by markets. Markets single mindedly aim to serve allocative efficiency. Optimal *allocation* is one thing; optimal *scale* is something else. (Daly, 1996: 32)

Campbell draws attention to this point in Chapter 3 in relation to the oil resource. It seems that some economists would rather dispute the scientific understanding of oil formation than acknowledge that their discipline does not adequately treat the problem of finite natural capital. So economic sustainability (sustaining the economy) is a very different matter from social or environmental sustainability and their reconciliation poses serious problems.

Urban transport and social sustainability

The naturalistic social analysis of the city by urban geographers such as McKenzie (cited above) was replaced in the 1970s in the foreground of scholarly thinking by Marxist class analysis (e.g. Harvey, 1973), only to be replaced in turn by a focus upon social networks as the key to understanding social structure (e.g. Soja, 1989). What has been termed 'reflexive modernisation' (Beck *et al.*, 1994) implies that social change can be accomplished through appropriate critique. Indeed the market economy can only be justified (in the utilitarian philosophy on which it is based) if its operation contributes maximally to the good of all.

Social theory thus postulates that society is not the outcome of the economy, economies are ultimately products of society. But we are frequently told in public policy documents that cities and nations have to adapt to 'globalization', as though globalization were a natural event beyond social control. Polanyi (1957 [1944]) and more recently, Hirst and Thompson (1996) have demonstrated by careful historical analysis that globalization is merely the continuing spatial expansion of market society driven by the governments of nation states. What then does 'social sustainability' mean when most of the key texts in social theory address not sustaining society but changing it?

Polèse and Stren (2000: 3) provide the clue. Social sustainability is defined as social integration: 'integrating diverse groups and cultural practices in a just and equitable fashion'. This is to say that *integration* is implicit in the whole idea of 'society', but there are better and worse ways in which integration can be achieved. One would not want to recommend the kind of forcible integration that occurs under authoritarian regimes. Hence it is necessary to add 'just and equitable'. These are not facts of a particular society but values to be achieved. Social sustainability, therefore, is about sustaining *progress* towards the kind of fair society in which the good of each (individually) coincides with the good of all (collectively). The reason that social sustainability must be considered separately from economic sustainability is that markets can disrupt, sometimes violently, the cohesion of society, and negate the principles of justice that hold society together without the constant use of force. What then is the relationship between economic and social sustainability, and between social and environmental sustainability? And how does transport figure in these relationships?

Economists (e.g. Sims, 1992; Rothengatter, 1994) claim that marginal cost pricing of transport infrastructure use is not unfair to poorer transport users: 'The rich run a higher mileage than the poor, so they will have to pay more if user charges are introduced, based on mileage travelled' (Rothengatter, 1994: 121). This is supposed to be an argument in favour of the 'progressive' effect of congestion charges. Under this sort of reasoning a 'flat' tax system in which all people pay the same proportion of their income in tax, would be regarded as progressive. This could only be true, however, if each dollar spent had the same value for everyone. But the extra dollar spent on travel by a poor person may mean the difference between having a job and not having one (i.e in some cases between starvation and survival). Whereas the extra dollar spent by a rich person may mean the difference between drinking a more or less superior brand of wine for dinner. Keen explains why economists mostly ignore the implications of this intuitively obvious fact: to do otherwise would make it impossible to claim, as economists do, that individual utilities can be summed to yeild social utility (Keen, 2001: 40–47). If, as Keen argues, they cannot, then economic efficiency cannot be equated with social sustainability. In fact a functioning market economy (which certainly has many of the virtues claimed by economists) depends on an integrated and just society that cannot be sustained by markets alone.

Banister (1994) in analysing the distributional effects of policies to internalize the social and environmental costs of transport implicitly acknowledges this enormous gap in conventional economics and discusses the complex realities of interspatial and intergenerational equity. On road use pricing he concludes:

> In affluent economies with high levels of car ownership, road-pricing is likely to lead to substantial sums being raised. Some road-users will be priced off the road, and the additional capacity may give benefits to those drivers continuing to use the road system until a new congested equilibrium is reached. The marginal low-income car owner will be the person to switch mode, destination, route, time or suppress the trip, whilst other users will choose to pay the charge. (Banister, 1994: 169)

Such a switch to alternative modes of transport may be desirable on environmental and health grounds. But, even in cities with an adequate alternative public transport system to switch to, the financial impact on poor and middle income groups is likely to be severe. Banister (1994: 170) suggests a 60 per cent rise in transport costs. In cities without adequate alternative transport the result would be socially disastrous and cannot be considered sustainable. Black (2000: 146) observes that while the rich will be able to buy the choice of continuing 'unsustainable' transport behaviour, lower income families will be the hardest hit by a move towards sustainable transport because they will be locked into current patterns. Mees (2000: 72) goes further: 'Given the difficulties with relatively inelastic demand, price setting, costing of the intangible and equity – problems for which [pricing] advocates have few answers and those generally risible – free market road pricing is unlikely to be the answer to urban travel problems'.

Is the environmental sustainability of cities *conditional upon* their social sustainability as argued by Polèse and Stren (2000: 15). Polèse (2000) observes that, in many of the case studies reported (in Polèse and Stren, 2000), public transport has increasingly been defunded in favour of road infrastructure spending; public transport is becoming the refuge of the poor and excluded: 'in short two systems of urban transport coexist: one for the owners of cars, one for everybody else. ... Two cities coexist in the same metropolitan area: the 'modern' city nurtured on the car; the 'other' city dependent on other, more traditional means of transport (including walking)' (320). This is precisely the picture that emerges in Chapter 10 of this volume. Rather tragically as Banerjee-Guha shows, in India arguments in favour of spending on roads are being misleadingly coupled with environmental sustainability to legitimate external consultants' advice (from the old paradigm) and redirect spending priorities from public transport to road infrastructure. At the same time a public transport solution that promises much better environmental outcomes and is consistent with both environmental sustainability and economic security is being ignored.

Rarely if ever are the opportunity costs of various forms of infrastructure investment considered. This is a matter of antiquated budgetary practice in developed countries. It is scandalous in poor ones. Road/car systems take up vastly more land and impose much greater social and environmental costs than fixed rail, cycling and walking. Yet in India, it seems, road systems are being built without any consideration of whether the community would rather have the money for flyovers spent on providing clean drinking water, safe sewerage infrastructure, education or health facilities. Even in Britain the wisdom of spending billions of pounds on new roads may be questioned when the public health system is on its knees.

If social sustainability need not coincide with economic sustainability, it may also conflict with the principle of environmental sustainability. From the viewpoint of 'modernity' the advantage of a city is that is that it brings a large number of people within rather close range of each other.[2] The advantage is both economic and social: a large market and a complex, culturally diverse and educated society marked by a certain freedom to doubt and challenge traditional norms. But once a city grows beyond a certain size the advantages of scale fall off rapidly – without transport. In the modern mega-city transport is necessary for people to take advantages of the social, and economic benefits it offers. Takeda and Mizuoka in Chapter 9 show how modern Japanese society, both urban and rural in interaction, was built around a certain form of public transport. Indeed it has been argued that Japanese prosperity depended on it (Hook, 1994, cited in Chapter 14 of this volume). Changing the way transport is provided, for economic or environmental reasons, has major distributional and social effects. Such changes may not lead to social breakdown but they are not 'socially sustainable' unless they sustain the progress of the society towards prosperity, freedom and justice for all and not just the entrenchment of class privilege. It is too strong a requirement, however, to insist that environmental sustainability is *conditional* upon such progress. Often the ignored environmentally sustainable solution is most consistent with social improvement and long term economic security.

Urban transport and environmental sustainability

Human societies and the economic systems they generate are, finally, subject to the limits of the 'natural' more-than-human environment. Environmental sustainability can be approached from a local and a global perspective. The local impacts of urban transport systems on human health are reasonably well known and some of these impacts are discussed in Chapter 7. Most of the negative local effects are associated with the mass use of private motor vehicles (both domestic and freight vehicles) on public roads. To a visitor from another planet it would seem astonishing that such a huge tally of human deaths and injuries from the crashes of private vehicle traffic is tolerated. Perhaps there is an underlying belief that such deaths and injuries are self-inflicted and thus avoidable, accompanied by the acceptance of the principle of individual responsibility, or, in language closer to truth but more foreign to modernity, acceptance of fate. Yet many of those who die or are injured have no choice at all in the matter. There are few citizens who do not over a lifetime have a close encounter with death at the hands of another driver.

Somewhat less well tolerated by modern society are the impacts on health and life of atmospheric pollutants. These pollutants are unavoidable in that nobody living in cities can opt out of contact with them unless they seal off their faces in gas masks to filter the air they breathe. This is not just a matter of ambient air quality in cities. Every traveller on a busy arterial road breathes the undispersed pollution from hundreds of motor vehicles on the road ahead. But pollutants mostly do not have an immediate effect and their precise longer term impact in small doses on health is not widely understood. Ultimately, the cost to human health may be as great as, or greater than, that of motor accidents.

There are also indirect health effects of motor vehicle use. The private car/public roads solution to transport problems has two systemic effects: it leads both to a spreading out of interacting land uses and the reduction of walking as a component of journeys between them. People have to travel further and they get less physical exercise doing so. This loss of physical exercise in the normal course of daily life can itself make people less healthy. The private car is particularly to blame because it provides a door to door travel option and its engine pollutes the atmosphere with noxious chemicals. This is quite different from public transport systems which can more readily accommodate electric powered, non-polluting vehicles (both rail and road systems) and in which there is always a need to walk short distances.

Finally, the door to door conception of car/road transport tends to demote urban public space to the status of an awkward gap between domestic space and the car. Yet it is also well understood that some of the finest achievements of the great cities are the squares and boulevards designed for people on foot. When public space is occupied by masses of parked and moving vehicles its environmental quality is destroyed. The movement to reclaim public space for people on foot has made great progress, particularly in Europe where the cultural value of public urban space is well appreciated.

Transport systems therefore create local environments of varying quality for their human occupants. But an exclusive focus on the environment of humans is regarded by many environmental philosophers as 'anthropocentric': that is, the only reason for valuing the human species above all others is that we are members of it (see Eckersley, 1992). As an ethical position that is unsatisfactory and probably does even not correspond with most people's ethical regard for non-human species (Low and Gleeson, 1998). Prudentially also we need non-human species in many known ways and many ways that are not yet discovered. The urban environment, and especially the peri-urban environment on the edge of great cities provides a habitat for many rare and endangered species. The car/road solution sprawls the city further and further into the peri-urban region, damaging the habitats of these species and threatening in some cases species extinction.

Hu, in Chapter 11 points out that a country as populous as China cannot afford to lose land with good food production potential. Chinese cities, which are often situated in naturally fertile regions, cannot therefore adopt the low density form of their American or Australian counterparts. Indirectly too, taking valuable agricultural land around cities for urban use puts additional pressure on more distant and more fragile ecosystems which may be forced into food production.

There is little doubt that preserving or improving the local environment of cities requires both a high level of service from a well integrated public transport system, less road traffic and, especially, reduced dependence on the private car. Even so, given the right sort of micro-climate that disperses photochemical smog, widespread and enthusiastic car ownership, and a basic level of public transport, car-dependent cities may still for the time being provide acceptable (if not particularly 'liveable' or socially just) urban environments for humans (for example the Australian cities discussed in Chapter 12 of this volume). But such acceptable cities are, nevertheless, far from environmentally sustainable. What citizens choose to do locally to the environmental quality of their cities is a different matter from 'environmental sustainability'. If the term is to mean something specific and different from 'environmental conservation' in general, *sustainability* has an irreducibly global dimension.

There are three global issues on which urban transport impinges: fuel, biodiversity and food, and the atmosphere. On the first issue, as Campbell points out in Chapter 3, the world is moving ever closer to a critical shortage of the one finite resource on which the world's transport systems, and therefore the production systems geared to them, depend: oil. This would hardly matter if there was an equally strong and far sighted effort under way world-wide to restructure transport systems away from fossil fuel dependence. Unfortunately there is not. There are technological advances in vehicle engines that would enable such a restructuring to occur. Destefani and Siores describe some of them in Chapter 5. But there are still major technical difficulties to be overcome and enormous costs in implementation. At present, in so far as they are conscious of the issue at all, the world's political and economic leadership seems to be relying entirely on the rising price of oil to stimulate restructuring over a relatively short period of time. Campbell concludes that the rising oil price is more likely to plunge the world into a global economic depression that will make the inevitable restructuring all the more painful and difficult. Will e-commerce be the solution? Stiller, in Chapter 4 does not conclude that either business to business or business to customer exchanges on the Internet are likely to lead to ecologically sustainable outcomes.

On the second issue, the earth's species do not belong to the nation in which they happen to be found but are part of a global human heritage. In less anthropocentric terms, the right of species not to be extinguished is a global right. 'Genocide' understood literally as the extinction of a species, whether by deliberation or neglect, may in future be regarded as an unforgiveable environmental crime. The spatial expansion of cities threatens both species habitats and food production for humans. The latter is beginning to be seen as problematic, especially for countries such as China and India which will have to feed huge populations from a diminishing supply of productive land as their cities grow. Brown (1999: 123) comments: 'The effects of the acute cropland scarcity emerging in some countries could affect many other areas of human activity. For example, it could fundamentally alter transportation policy, favoring the development of more land-efficient bicycle-rail transport systems at the expense of the automobile'.

The third issue is arguably the most serious and the least likely to be resolved peacefully and justly. It is also the most critically important for environmental sustainability. Global warming is happening because of the growth of the world's carbon based economy during the last two hundred years. There is an extraordinary degree of consensus on this. Very few scientists disagree, along with some commentators who tend always to cite the same sources. The Intergovernmental Panel on Climate Change (http://www.ipcc.ch/), its thousands of advising climatologists, and the Meteorological Offices of the USA, Britain (The Hadley Centre: http://www.met-office.gov.uk/research/hadleycentre/), and most other European countries believe that their research tells them that the Earth *is* warming and that human economic activity is responsible.

Global warming will affect just about every aspect of life on the planet both human and non-human. Many species will become extinct as their habitat is destroyed by the climate changing too rapidly for them to adapt. Coastal cities and low-lying regions (e.g. Bangladesh with 124 million people) will be threatened with rising sea levels. Climate change is perhaps the first truly global problem of environmental sustainability that human society has ever had to face. The world is at present ill equipped to do so. That the problem is deadly serious can be gauged from the following figures.

Climate change: business as usual

What will happen if we go on as we are? World emissions from burning fossil fuel in 1999 were 6143 billion tons (International Energy Agency, 1999). The Intergovernmental Panel on Climate Change has concluded that to stabilise the Earth's climate system we must stabilise the amount of greenhouse gas in the atmosphere at about 450 parts per million of volume (ppmv). To achieve that figure we must in this century reduce carbon emissions by about 60 per cent. In terms of CO_2 emissions from burning fossil fuel that means getting down to about 2,500 billion tons per year. Figures published by the International Energy Agency (http://www.iea.org/) show that, even on the most optimistic scenario, China and India together will be emitting more than 3,800 billion tons of CO_2 within 50 years. This is certain to be an underestimate unless economic growth collapses in China, since it is based on the assumption that China will continue to reduce its fossil fuel dependence at the same rate as the last ten years (1989–99) when much of the old coal burning industry was being phased out.

A more likely – though still optimistic – scenario is that in which (by 2049) most developing nations can expect to reach the rate of fossil fuel dependence of the USA today, while the US continues to reduce its dependence at the same rate (1989–99), with rates of economic growth of the same period continuing. Under that scenario, China will be emitting 22,000 billion tons, India 1,360 billion tons, and the USA 2,904 billion tons, totalling 26,264 billion tons. However, if China, India and the USA continue to grow economically at the rate they did between 1989 and 1999 but do *not* improve their carbon dependence much beyond that of today, then within 50 years their emissions will grow to a staggering 92,000 billion tons, nearly forty times what is required to stabilise the climate, and for just three nations.

The problem is not Chinese cities or Indian cities as they are today. American and European cities are much more problematic in their carbon emissions. The problem is the continuing growth of fossil fuelled energy consumption in North America, Europe and Australia and the wholesale transfer of the so called 'developed' way of life (the 'old paradigm' in both American and European variants), including its appetite for mobility, to cities of the developing world.

There are also many factors that will act in the next fifty years to alter projections. One is the rising price of oil (Chapter 3). But it is not a happy thought that the surest route to transport sustainability lies in global economic depression. Another is, as Hu describes in Chapter 11, that the deteriorating local environmental quality in mega-cities (in China) and the shortage of land for agriculture, coupled with tolerance of high population density and strong governmental regulatory capacity will simply rule out the pursuit of the car/roads solution. However, avoidance of such an urban disaster is hardly a foregone conclusion. Barter, Kenworthy and Laube in Chapter 15 note that both India and China appear to be in the early stages of a boom in motorisation which, if it continues over many years, will not only ruin the local environment of cities in those countries but threaten the global atmosphere.

In the USA transport rhetoric emphasizes sustainability, and there are signs of significant change in values at grassroots level (see Chapters 2 and 6). But investment continues to flow into infrastructure schemes: mostly new road projects. The state Departments of Transportation (state DoTs) still have much power to decide how funding for transportation is spent. For example, the Interregional Transportation Strategic Plan for California makes clear where its priorities lie: 'To protect and realize the maximum benefit from our investment in the highway system, we must continuously maintain and rehabilitate it…The state highway system supports, directly and indirectly, the state's economy and its continuing growth' (Sels, 1998: 15). Meeting this goal entails vast investment not only in maintenance, but also in enlarging and upgrading the state's highway system. The New York Metropolitan Transportation Council espouses the goal of sustainable mobility but envisages a potential US$65 billion wish list of transport investment projects (NYMTC, 2000: 2). The road lobby is very powerful: the American Highway Users Alliance wants to force states to use 85 per cent of federal funding for highway projects and allow states to use the remainder for other projects at state DoTs discretion.

The fast rail transport developments proposed for Europe add a further dimension to the problem: that of networked cities in which movement

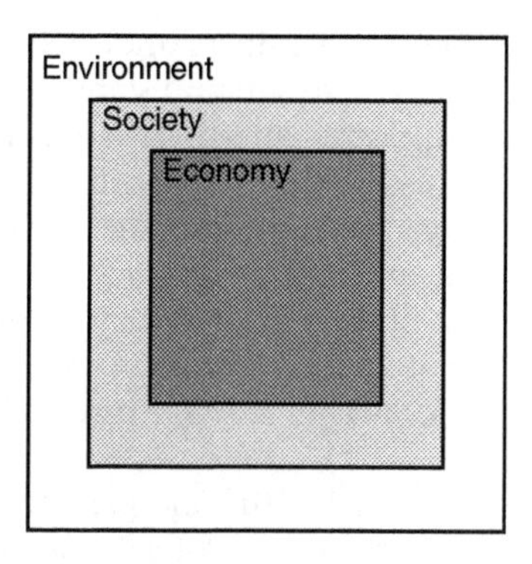

Figure 1.2 Sustainability as a set of nested boxes

between cities escalates dramatically and entire regions are mobilized. While better public transport is necessary to improve local environmental quality, and is particularly needed in cities of the developing world, the huge investment in high speed trains in Europe does nothing for environmental sustainability at global level and will merely increase greenhouse emissions. What is needed, as Whitelegg argues (Chapters 7 and 14) is the management of demand to reduce unnecessary travel.

From the above discussion it will be clear that sustainability should not be portrayed as the outcome of the classic engineering triangle of forces, but rather as the outcome of a set of nested boxes: economic sustainability inside social sustainabilty inside environmental sustainability (Figure 1.2). This perspective does not imply that society and economy have to be sacrificed to the environment, merely that what goes on in society and economy is subject to the natural environment which supplies the inputs and absorbs the wastes. Also that what goes on in the economy is subject to the fairness, integrity and stability of society. The city, after all, is a geographical system that exploits the environment for human purposes. What has to be planned is responsible use of the environment in a way that is socially just and can continue indefinitely without running down the environment's capacity to provide inputs and absorb wastes, and without destroying the complexity of life evolved over millions of years.

Ecosocialization

If change to the world's transport systems to make them more sustainable is going to happen, how will it happen? Nothing less than a 'paradigm shift' is entailed. Thomas Kuhn (1962), and his successors in the sociology of science such as Bruno Latour (1987) have shown that what counts as acceptable knowledge is embedded in the society of those possessing it, and certain key assumptions determine how its members understand the object of knowledge. Such key assumptions, (or 'paradigm') stand until they are replaced by an alternative set of assumptions – and there is a 'paradigm *shift*'.

It is easy to see that the assumptions supporting current transport policy are deeply embedded not only in society at large but more specifically in the policy communities inside and outside government that create infrastructure and transport. This is shown most clearly in the Australian case discussed in Chapter 12, but it is also implicit in many of the other national studies. Following the work of Hajer (1995) we can call these sets of assumptions 'storylines'.

Each of the storylines comprising the 'old paradigm' seems like common sense. Each is embedded in practices and institutions: bureaucracies, dominant professions, transport planning agencies, customary methods of funding infrastructure and apportioning costs (environmental and social as well as individual and governmental), media treatment of transport, advertising, production corporations, transport products and technologies, trade unions, highway service organisations, class power (which did not disappear even though marxism is unfashionable), racial and gender inscriptions, and a vast array of regulations. There is no conscious conspiracy, but providers of transport speak with a common vocabulary to define problems and create solutions. This is neither surprising nor evil but change cannot be postponed.

Some of the storylines, institutions and powers are revealed and challenged in this book. If there is to be a change of paradigm, common sense will have to be re-evaluated. How can *freedom* of movement be protected while *forced* movement (e.g. the inescapable commuter trip) is reduced? What about freedom *not* to move? What sort of cities do we want and what sort of transport system would deliver them? How can we plan and manage a transport system as a mutually supporting whole? What is sacrificed to permit unimpeded vehicle movement, and is unimpeded vehicle movement healthy for people? Does infrastructure bring rapidly diminishing economic returns – especially when the hidden costs are accounted for?

The paradigm shift will not come easily or quickly. But if it is true that the major technological systems of the world economy are severely eroding the environmental resource base of the planet, the 'natural capital' in Daly's terms, then we need to consider a much longer term and even more profound process of societal change. We have elsewhere termed this process 'ecosocialization', drawing on the seminal work of Karl Polanyi (Low and Gleeson, 2001; Low, 2002). Polanyi described the social countermovement evolving over a hundred years or more in England to protect society against the destructive effects of expansion of the market promoted by national governments. The various institutions of transport (including public roads) are among the social supports resulting from that movement that today make life for humans in cities tolerable. They represent a trade-off between the market and society – but at cost to the environment.

The environmental crisis has provoked a *new* countermovement on the part of society to protect its ecological supports and adapt capitalism to its ecological limits (see Bernard, 1997). We term this movement 'ecosocialization'. It is not a class movement (socialism), nor a 'social movement' like the 'anti-war' or 'pro-life' movement, though of course it may mobilize communities and groups into action. Ecosocialization springs up within society's institutions and organizations as a new common sense (paradigm) establishes itself. New questions are posed and answered. New values superimpose themselves over the old. Existing institutions change their behaviour. New rules and procedures are established to implement the new common sense. New technologies are gradually embedded to deliver the new values. Ecosocialization is a change at the level of the cells and organs that make up the body of society. The 'eco' prefix signifies an awareness that the environment is at stake and must be protected in future human development.

Is ecosocialization taking place and what does it entail? Newman in Chapter 2 points out that changing social values driven by changing demographics are leaning to a more urban and less suburban lifestyle. His focus is the USA and Australia. This is reinforced by a changing nexus between the 'knowledge economy', communication technology and concern for sustainability. His focus (with an emphasis on the USA and Australia) is pressure for ecosocialization welling up from below. Unquestionably, also, automotive technology will be part of the solution to both global warming and the peak of oil but it cannot do much about the health effects of dispersed production and consumption. The rising price of fuel will accelerate technological change. But the difficulties should not be underestimated. As is evident from Chapters 4 and 5, technology is not a magic wand. All technologies involve difficulties, costs and contradictions: the popular beliefs that it will one day be possible to run cars on water or that economic interactions will mainly be conducted electronically are nothing more than myths.

The picture presented in Chapter 7 is dark – it may be too dark for some readers to accept readily, yet it is backed by evidence. Whitelegg depicts a developed world (Europe is the focus here) in which a health and environmental catastrophe is unfolding. Society is being spatially reorganized by central authority around a dispersed logistical model of production and distribution, the costs of which are are being continuously loaded on to public ill health, social injustice and environmental destruction. Vast sums are being spent to facilitate unnecessary movement of people and products. The failure to deal with the unintended effects of transport in the modern city today, he argues, is equivalent to the failure to deal with the unintended effects of defecation in the city of the Industrial Revolution. The latter caused plagues of cholera and typhoid, the former is causing epidemics of death and disease every bit as severe. In many ways transport-generated urban dysfunction is more severe, as well as more subtle, because of the global threat of climate change caused by the locally innocuous carbon dioxide emitted from the transport sector.

In Europe ecosocialization is proceeding slowly and can be found at government level in policy and rhetoric, but it is not yet influencing the crucial actions which cause the distance-intensive society and generate the social and environmental costs of transport. This is probably because the full impact of transport is only just beginning to be appreciated in central policy-making circles in terms of long term sustainability rather than short term economic growth. It seems that there are severe barriers to change created by existing paths of thought that are preventing paradigmatic shift.

Tengström's policy analysis in Chapter 8 of what can be regarded as the world's vanguard nations in transport sustainability (Denmark, the Netherlands and Sweden) provides a nuanced perspective on social change. He shows how, and considers why, different nations proceed at different rates and at different times with the change demanded by 'sustainability'. Reflecting on his observations he introduces a number of key concepts to explain the variance in policy change: the secular upswings and downturns of attention on political issues, or 'issue cycles' – the perception of crisis seems to play a part here; the discourse theoretic concept of storyline ('sustainability' and countervailing stories supportive of conventional transport planning); ecological modernisation (with the Netherlands as the outstanding model); varying political cultures and processes and geographical differences; government failure and acceptance failure (on the part of citizens). Tengström's analysis and evidence does not contradict the picture presented by Whitelegg but provides an explanation and also a way forward via the simple but neglected process (even in vanguard countries!) of real communication and dialogue between citizens and government.

Can we draw some hope from the experience of different regions? It is evident that change is gradually occurring in the United States – people are simply demanding something different from the car dependent non-place urban realms that American suburbs have become. The change is being driven from the bottom up (Chapters 2 and 6). In Europe there is a concerted push coming mainly from the neighbourhoods, the cities and their planners but finding expression and financial support at the level of the European Union. The range of policies is discussed in Chapter 13. There have been some striking success stories in demand management which show that economic growth can be decoupled from growth in movement (Chapter 14). Infrastructure investment policy, however, both in the USA and Europe remains fixed on maximising mobility and expanding the space of economic exchange under the assumptions of the old paradigm. It is grossly unsustainable and highly resistant to ecosocialization.

Asian cities face a critical choice. The varying experiences of cities of this vast region demonstrate that wealth and motorization do not correspond.

There are considerable differences in policy orientation, the more successful and cost-effective cities investing heavily in public transport as mode of choice and not allowing that mode to become a residual welfare service for the poor. The evidence presented strongly suggests that an early decision to prioritize public transport over private transport brings long-term benefits not only in environmental and social sustainability and improved public health but also in economic security (Chapter 15). Yet the lure of the old paradigm of growth and mobility American style remains powerful. It has been disseminated and promoted in Asia and India by consultants from the developed world (mostly from the USA) and it will take a long time and much continuing effort to overcome. Important is the development of critical, analytical and problem-solving capacity within the nations and cities themselves so that these cities can address their own problems with locally generated rather than imported solutions.

Conclusions

Cities gather people together for the purposes of social and economic interaction. But with increasing scale this gathering leads either to congestion – with too little space for living, or separation – with too much time spent in conducting exchanges over space. For a hundred years transport based on burning carbon fuels deposited in earlier periods of extreme global warming (millions of years ago) bridged the gap between congestion and separation. Extending personal mobility to the people was the freedom offered, and mobility and its symbols continue to exert a powerful hold over public perceptions. The technology of the motor car and the road, whatever one thinks of its social and environmental consequences, is a wonder of contemporary culture.

The bridging function of transport was not carried out without cost. The transport solution to population congestion led inevitably to congestion of the transport corridors. But the cost could be accepted because it was either placed on society as a whole in the form of massive expenditures on transport infrastructure, land and health, or distributed to the more powerless sections of society, or loaded on to the environment in the form of localized air and water pollution and the use of the whole atmosphere as a carbon sink. The result has been the growth of an extensive and highly sophisticated *institutional* system supporting the production, design and financing of transport infrastructure built on the assumption that fossil-fuelled mobility could continue for ever.

Urban transport today is not sustainable for reasons that will be developed fully in the chapters to follow. The situation is extremely serious and requires immediate action by governments, communities and businesses. The time of fossil fuel is rapidly coming to an end both because of the environmental consequences of returning fossil carbon to the atmosphere and because the fuel itself is running out. Increasingly mobility will become a zero sum game. Mobility for some will be at the expense of immobility and disease for others. There will be distributional effects both within and between places (nations, regions, cities). Cities that have invested in diverse alternatives, especially in a well planned and high quality service offered by mass transit systems, will be in the best position to survive and prosper both socially and economically. Cities that are dependent on a single mode of transport, especially one based on private vehicles and roads, will face growing hardship, pollution and social strife.

Technological change will undoubtedly provide some solutions. But we must never be tempted to believe that technology will solve what are essentially problems of planning and management. Public expectations and demands may also have to change. But the biggest 'roadblock' to ecosocialization is likely to be the inertia of the institutional system and its beliefs which remain devoted to the pursuit of a worn out paradigm.

Notes

1. This is a longstanding line of argument. H.F. Storch in his 'Course in Political Economy' of 1823 writes: 'Every detour, delay, intermediate exchange which is not absolutely necessary for this purpose, or which does not contribute to diminishing the circulation costs, harms the national wealth by uselessly raising the prices of commodities' (cited by Marx, 1973: 636).
2. Many sociologists (since Max Weber) have observed that the growth of cities is associated with the advance of 'modernity' and its corresponding challenges to traditional forms of social integration. With such challenges, modernity has also brought certain freedoms and a relaxing of social control. We can follow van der Pijl's characterisation of the dominant form of modern polity as 'Lockean' after the English Enlightenment philosopher, John Locke. The Lockean form admits of a wide range of variation from the social democracy of Sweden to the market society of the USA, from the state corporatism of Singapore to the consultative politics of the Netherlands, embracing republican, federal and monarchical constitutions. But modernity also brings new expressions of class power and new ideological structures associated with the 'individual', 'private property' 'economic rationality', the 'rule of law' and the market. Of course one may reasonably question whether such 'modernity' is desirable. But the fact is that states all over the world from Brazil to Bangladesh, and South Africa to China have embraced modernity and fashioned their societies to a greater of lesser degree around the principles of Lockean politics.

Part I

Global Issues in Transport Sustainability

2

Global Cities, Transport, Energy and the Future: Will Ecosocialization Reverse the Historic Trends?

Peter Newman

Introduction

A simple view of transport and the future suggests that mobility has always increased with wealth (see Rainbow and Tan, 1993; Schafer and Victor, 1997). Thus as wealth grows it is anticipated that there will be more and more car use and that the other transport modes will all reduce. Yet a recent survey in the wealthiest economy in the world has suggested that 98 per cent of US citizens want to see public transport improved and 68 per cent want to see public funds spent on rail systems. The survey included the Chicago area, and most of those surveyed – almost 75 per cent – were from suburbs or outside a central city; 89 per cent agreed that traffic had worsened nationwide and two-thirds of the sample wanted to see high speed rail as a result (APTA, 2000). Where does such sentiment come from? Is it just romantic and ill-informed? Or is it suggesting that some new limits to the growth in car use could be approaching? Is there an 'Ecosocialization' trend underway that could decouple transport mobility and wealth?

This chapter will show that understanding the future of transport (and fuels) is not just a matter of projecting trends. It requires an understanding of cities and the social values that shape them. Social values shape transport technology, they shape the emerging knowledge economy and the demographic forces of our era, they shape urban governance and they shape the consumer urban lifestyle market for housing, land, location and transport mode (see Figure 2.1). In this chapter some emerging trends will be pointed out that could indicate how changing social values may indeed be pushing cities into different pathways that could be significantly different in their transport outcomes from simple trend projections. We can term these 'ecosocial' values.

In particular, the guiding question will be asked of a number of overlapping influences in city development: Are they making the city more car-dependent or less? Are these forces making cities more or less urban? The

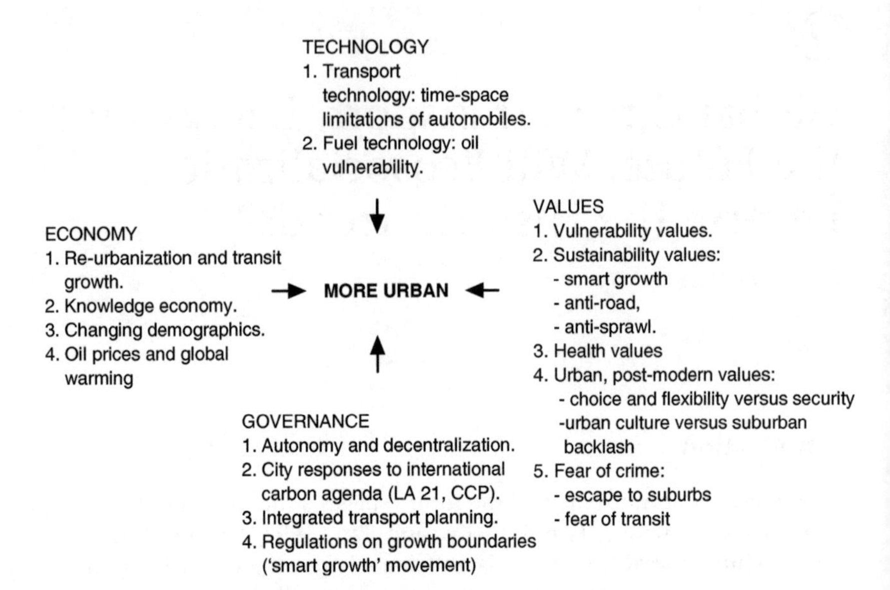

Figure 2.1 Key influences on city development

consequence of this answer directly influences car-dependence. If a city is becoming more *urban* it is more than likely going to be less *car-dependent* and if less urban then it would be likely to become more car-dependent. Thus this chapter will provide an outline of some key influences on city development and in particular how they relate to the increase or decrease in urban rather than suburban values.

Car-dependence and technological forces

In 1989 we defined car-dependence as building a city with the assumption and priorities that most people will drive cars.[1] We suggested that this was reflected in gross densities under 30 people per hectare and transport infrastructure dominated by roads. Very low densities mean few options for walking, bicycling or public transport due to the sheer distances involved unless a very long time is taken to make the trip. However density is not everything. In very dense cities it is still possible to be car-dependent unless other options are available. Walking will always be easier, though perhaps quite dangerous, but buses can be just as slow as in sprawling suburbs, especially in the third world, because buses get stuck in traffic and again only cars offer a travel solution within a realistic timeframe. Car-dependence can thus be generalized in terms of a thirty minute average journey to work: situations where a thirty minute average journey to work *is attainable only by car*. The significance of the thirty minute limit becomes apparent when we consider the impact of transport technology on the shaping of cities.

Transport technology

The twentieth century has been dominated by the growth of automobiles and they have shaped the land use of cities, especially in wealthier developed nations. In Newman and Kenworthy (1999) it was shown how cities based on walking speed spread out from the centre for no more than 5–8 kilometres, and cities based on nineteenth century public transport ('transit') technology spread in corridors for 20–30 kilometres. Cities based on the automobile can spread over 50 kilometres in every direction. This era is reaching some limitations due to the sheer time and space constraints of the transport technology. The basis of this technological limit can be understood with reference to a fundamental social value which people strive to attain in every city. In all cities today and throughout history there is a pattern called the 'constant travel time budget'. This is sometimes referred to as the Marchetti constant. On average people do not like to travel for more than one hour for a normal day's activities. Thus the average journey to work is half an hour whether it is in *walking cities, transit cities*, or *automobile cities*. Urban land use adapts to the infrastructure provided to enable this pattern to occur.

The limits to each technology of transport were overcome by a new technology that enabled successive phases of city-building and economic development. The three city types (*walking, transit, automobile*) can be related to the last three Kondratiev business cycles with their characteristic transport technologies. Analysts who examine technology in history show that technological change both changes the way we live and work and is also an expression of our fundamental values (Metcalfe, 1990; Freeman, 1996; Linstone and Mistroff, 1994; Freeman and Soete, 1997). The phases outlined above can be seen to have certain key technologies in each phase: the first stages of industrialization were based on cotton and iron, the next phase on coal, then steel and finally oil (see Table 2.1). However the technologies were also associated with different ways of doing the production process and different social values.

The fourth cycle which reached its climax after the Second World War was based on simple mass production lines – involving essentially linear thinking. This form of production is sometimes called 'Fordism' after Henry Ford who made the first industrial production line in his Detroit car factory. The phase we are now entering, the fifth cycle, is clearly related to the microchip, to information technology and how it can be used to control production and transfer knowledge and services. But the fifth cycle is also about how we are becoming a society more based on networks and multiple goals. Part of the new paradigm is 'sustainability'. This is setting the new direction for how all

Table 2.1 The five cycles of economic activity

Long wave	Time period	Main industries	Key factor	Business paradigm
1	*1770s to 1840s* Industrial Revolution to 'Hard times'	*Early mechanization* Textiles, potteries, canals *Steam power and railways*	Cotton and iron	Capital-based local industries
2	*1830s to 1890s* 'Victorian Prosperity' to 'Great Depression'	*Trains, steamships, machine tools*	Coal	Large firms
3	*1880s to 1940s* 'Belle Epoque' to 'Great Depression'	Electrical and heavy engineering Electricity, cable and wire, trams, radio	Steel	Giant firms, monopoly, oligopoly
4	*1930s to 1990s* 'Golden Age' of growth and full employment to 'Structural Adjustment' crisis	*Fordist mass production Cars, trucks, tractors, aircraft, petrochemicals, fertilizers*	Oil	Multinational firms, subcontracting, hierarchical control
5	*Late twentieth century* 'Global Recession' to Next wave of economic activity	*Information technology Environmental technology Sustainable transportation*	Communication and control systems	Networking systems, flexible specialization, community scale

Source: Freeman and Perez (1988), *passim*.

forms of technology and social interaction progress. It is the ultimate approach to 'network' thinking rather than 'linear' thinking as it integrates not only natural capital and financial capital but also social capital (Putnam, 1993; Selman, 1996). Transport is a key technology. The economy is not formed by its mode of transport but transport develops in close connection to the new phases of technology and is also an expression of our fundamental social values. It is not hard to see how transport has developed in its broad patterns due to the large economic cycles outlined above; but it is also possible to see how different social values are expressed and prioritised in the

Table 2.2 Transport technology advances and their links to economic cycles

Economic cycle	Major intercity transport technology	Major city transport technology
1770s–1840s	Ships and horsedrawn carriages	Walking
1830s–1890s	Steam trains	Trains
1880s–1940s	Steam trains and Electric trains	Trams
1930s–1990s	Trucks and planes	Cars

combinations of various transport technologies. The broad patterns in terms of transport technology can be summarized in Table 2.2.

The limits on the extent of car-dependent cities are now being felt. Urban sprawl is now going beyond the ability of people to reach most major destinations in half an hour. The new limit is not due to the cars themselves but to the infrastructure for moving them. Highways can only carry 2,500 people per hour in cars on a single lane, and the more these lanes are built the more they destroy the very qualities of *liveability* in a city that the accessibility is trying to provide. The number of cars that can be squeezed into the space created for high capacity road infrastructure is now reaching social limits – space can only be made for more cars and trucks if cities begin double-decking freeways. The limits are experienced as 'environmental' limits and are sometimes dismissed as just 'values' but they are real expressions of deeper limits to the technology. The appeal of transit is that a bus lane can carry 7,000 people per hour and a train can carry 50,000 people per hour on the same space as one lane of traffic (2,500 people). Thus a vast majority of Americans do not see freeways as a solution to the transport problem any more. They want to see transit solutions and they want to see urban growth boundaries imposed (see below data on 'smart growth') so that travel distances are not so great.

These time-space limitations are being experienced even more rapidly in dense cities in the developing world and are the main reason why these cities will never have the car ownership/car-dependence of cities that developed with the car in the twentieth century (see Chapter 11 this volume). The key issue for the future is whether the new knowledge economy is going to alter cities and, through information-communication technology (ICT), provide the basis for resolving the time-space dilemma confronting car-dependent cities, or whether a new kind of technology will emerge that can overcome these limits. The link to the knowledge economy is outlined below. At this stage the trends suggest that ICT–sustainability–knowledge economy interactions could see a return to a less car-dependent city. However this will also depend on the way that social and political values and priorities are played out in different cities in different cultures.

Fuel technology

The limits are also being seen in the global limits on oil which have fed the growth of cities in the past 50 years. The limits may be as much perceptual and political as they are technical. Cities are traumatised by oil shocks and their sense of vulnerability spurs decisions to favour a greater diversity in fuel base. Data from our global cities study show 44 cities with significant variations in their vulnerability to oil shock (see Table 2.3).

Table 2.3 Transportation energy use per capita in global sample of cities, 1990

City	Private transportation			Public transportation			Total trans. energy (MJ)	Total trans. energy/$ of GRP (MJ/$)
	Gasoline (MJ)	Diesel (MJ)	% private of total	Diesel (MJ)	Elec. (MJ)	% public of total		
Sacramento	65,351	10,998	100	305	19	<1	76,673	?
Houston	63,800	7,325	99	499	0	1	71,624	2.74
San Diego	61,004	5,689	99	527	28	1	67,248	?
Phoenix	59,832	4,507	100	301	0	<1	64,641	3.14
San Francisco	58,493	6,187	98	935	275	2	65,890	2.12
Portland	57,699	12,358	99	614	27	1	70,698	?
Denver	56,132	11,560	99	594	0	1	68,286	2.78
Los Angeles	55,246	6,279	99	643	0	1	62,167	2.50
Detroit	54,817	7,522	99	405	0	1	62,744	2.78
Boston	50,617	6,676	98	845	252	2	58,391	2.10
Washington	49,593	9,732	98	753	376	2	60,454	1.68
Chicago	46,498	8,355	98	1,060	208	2	56,121	2.16
New York	46,409	3,747	97	975	494	3	51,626	1.80
American average	**55,807**	**7,764**	**99**	**650**	**129**	**1**	**64,351**	**2.38**
Canberra	40,699	3,333	98	962	0	2	44,995	?
Perth	34,579	5,965	98	851	0	2	41,395	2.34
Brisbane	31,290	7,071	98	632	284	2	39,277	2.10
Melbourne	33,527	4,613	98	411	338	2	38,890	1.84
Adelaide	31,784	4,359	97	953	6	3	37,103	1.88
Sydney	29,491	4,481	97	776	326	3	35,074	1.63
Australian average	**33,562**	**4,970**	**98**	**764**	**159**	**2**	**39,456**	**1.96**
Calgary	35,684	10,535	98	808	106	2	47,133	?
Winnipeg	32,018	6,358	97	989	0	3	39,366	?
Edmonton	31,848	11,116	98	1,027	69	2	44,060	?
Vancouver	31,544	4,740	98	743	184	2	37,211	?
Toronto	30,746	1,058	95	1,286	523	5	33,613	1.49
Montreal	27,706	?	?	1,019	261	?	?	?
Ottawa	26,705	5,421	95	1,526	0	5	33,562	?
Canadian average	**30,893**	**6,538**	**97**	**1,057**	**163**	**3**	**39,173**	**?**
Frankfurt	24,779	12,771	98	243	499	2	38,293	1.09
Brussels	21,080	6,297	95	635	883	5	28,895	0.96
Hamburg	20,344	15,463	98	556	352	2	36,716	1.21

Table 2.3 Continued

City	Private transportation			Public transportation			Total trans. energy (MJ)	Total trans. energy/$ of GRP (MJ/$)
	Gasoline (MJ)	Diesel (MJ)	% private of total	Diesel (MJ)	Elec. (MJ)	% public of total		
Zurich	19,947	3,875	94	609	813	6	25,244	0.56
Stockholm	18,362	6,636	93	1,068	751	7	26,817	0.81
Vienna	14,990	4,387	94	538	689	6	20,603	0.74
Copenhagen	14,609	4,091	92	1,313	372	8	20,385	0.68
Paris	14,269	9,026	96	323	946	4	24,241	0.72
Munich	14,224	2,598	92	210	1,166	8	18,197	0.50
Amsterdam	13,915	5,096	96	456	375	4	19,843	0.79
London	12,884	9,140	94	693	657	6	23,374	1.05
European average	**17,218**	**7,216**	**95**	**604**	**653**	**5**	**25,692**	**0.83**
Kuala Lumpur	11,643	7,600	96	774	0	4	20,017	4.92
Singapore	11,383	4,957	90	1,608	131	10	18,079	1.40
Tokyo	8,015	9,305	95	212	711	5	18,243	0.49
Bangkok	7,742	7,409	83	3,026	0	17	18,176	4.75
Seoul	5,293	2,604	82	1,551	168	18	9,615	1.62
Jakarta	4,787	3,845	95	440	0	5	9,072	6.02
Manila	2,896	2,734	77	1,698	8	23	7,335	6.67
Surabaya	2,633	2,684	95	294	0	5	5,611	7.73
Hong Kong	2,406	5,679	84	1,217	310	16	9,612	0.68
Asian average	**6,311**	**5,202**	**89**	**1,202**	**148**	**11**	**12,862**	**3.81**

Note: The cities for which no energy per unit of GRP is available are those cities where we do not have the GRP data.

The data show that fuel use is not obviously related to wealth. Some wealthy cities in Europe and Asia have much less transport fuel use than cities in the US and Australia. The data indicate that a major middle eastern conflict resulting in an oil shock would have significant economic impacts on the most vulnerable US cities like Phoenix and Houston. These cities may however choose to take that risk. However, with the balance of oil production becoming further concentrated in politically sensitive parts of the world (see Chapter 3 of this volume), it is likely that many cities will choose to become less vulnerable. There is evidence to suggest increasing use of strategies to reduce oil vulnerability in cities like Toronto, Freiberg, Portland and Curitiba. Many cities have merged this strategy into their consideration of how to deal with greenhouse gas issues. There are now over 400 cities in the ICLEI (International Council for Local Environmental Initiatives) *Cities for Climate Protection* program (http://www.iclei.org/co2/).

The issue is also not just one of vehicle or fuel technology either, as is shown by the data on the Australian fleet fuel efficiency over the past 35 years. There have been no gains in fuel efficiency (Figure 2.2). This is largely due to social

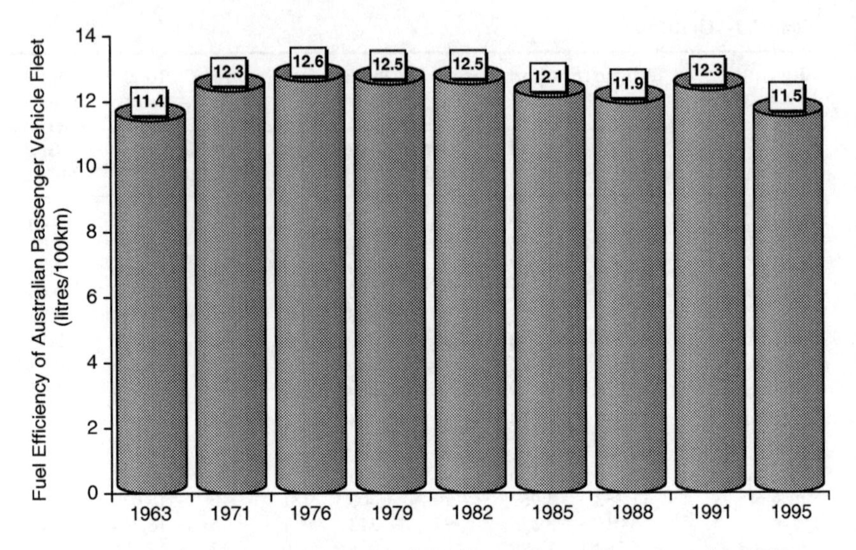

Figure 2.2 The fuel consumption rate of the Australian car fleet since the 1960s
Source: Motor Vehicle Usage Surveys summarized in Mees (2000).

factors such as the extent to which poverty means that older vehicles continue to be part of the vehicle fleet when no other suitable transit option is available. In Australian cities the poor are increasingly the most car-dependent and hence the vehicle fleet is now one of the oldest in the developed world. The social values associated with the move to sports utility vehicles (SUVs) have also contributed to the world-wide decline in fleet fuel efficiency. The overarching social value with respect to transport technology, however, is how the city will change with regard to the new economy and how this may solve the limits now being researched in time and space due to automobility.

The economy of cities

The city is profoundly affected by transport priorities – as with other infrastructure networks. The nineteenth century cities of the industrial world were rapidly urbanised without physical infrastructure. By the end of the 1890s there was an environmental and moral problem that demanded a solution. The solution came in the form of infrastructure – for transit, sewage, water, power, and the provision of suburbs. The twentieth century was a century of infrastructure and suburban development. The motto of the Town and Country Planning Association 'nothing gained by overcrowding' has dominated physical planning, especially in the English speaking world. The result was car-dependence, either planned or because it was an outcome of other economic forces. The question today is: where will the next trend in urban development take us as the Fifth Kondratiev Cycle sets in?

It is suggested in this chapter that there are new constraints and new influences which are different from those that shaped the cities of the twentieth century. All of them, if examined carefully, can be shown to fit a pattern where it could be expected that car-dependence will reduce. The two key economic factors are the new demography of cities and the knowledge economy.

Demographic factors and 'reurbanization'

The simultaneous operation of four demographic factors has resulted in reduced size of households in most developed Western cities and, correspondingly, a different set of priorities from the twentieth century one of providing suburban houses for families. These factors are the rapid increase of women in the workforce, the decline in households having children (now only one in five in most developed cities), the ageing of the population, and the increase in the number of one and two person households (now over 50 per cent in most developed cities). None of these factors individually means that people will consciously want to get rid of their cars and become transit users, but they will more than likely want to be more urban and not less urban because that is where the services are. Suburbs are mainly good for child rearing and there is no shortage of these for the rapidly declining family house. Hence, around the developed world there has been a rush in the development of more urban-oriented housing closer to urban services. This usually means housing closer to urban transit and shorter journeys of all modes. Hence, a less car-dependent lifestyle is growing again.

The above process can be described as 'reurbanization' as opposed to suburbanization – development on the urban fringe at low density. Evidence of this process is apparent in all the global cities for which we collect transport and land use data. In our book *Sustainability and Cities* we gave early evidence of a simultaneous trend to increased density and increased transit with a slowing down in car use. Data can now be shown for Australian and US cities in the 1990s that demonstrate these trends (see Figures 2.3–2.6). The figures are set out below and indicate that transit growth is now around five per cent per year in US and some Australian cities, and car use growth is much less, indeed it is in decline in some cities on a per capita basis. This process can also be related to the growth in the knowledge economy.

The knowledge economy

The knowledge economy is bringing 'urban' cities back in again because of the importance of face-to-face interactions in the creative development of the economy. Early notions about information technology suggested that its impact on cities would be to create 'community without propinquity', to disperse people into 'non-place urban realms' or exurbs, where distance would be overcome electronically and people only needed to 'telecommute'

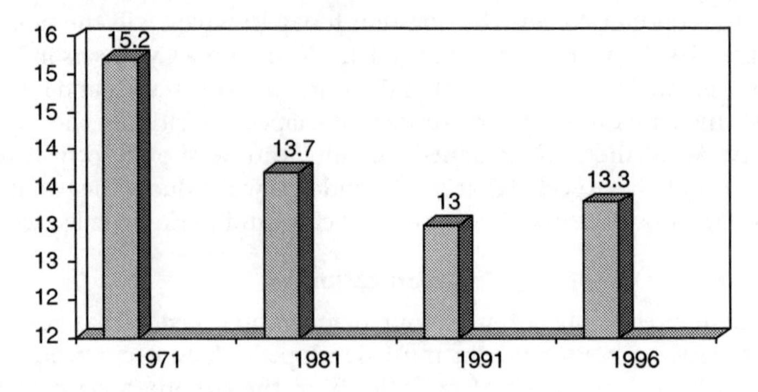

Figure 2.3 Density trends, 1971–96: Melbourne, Perth, Brisbane, Sydney (people per hectare)

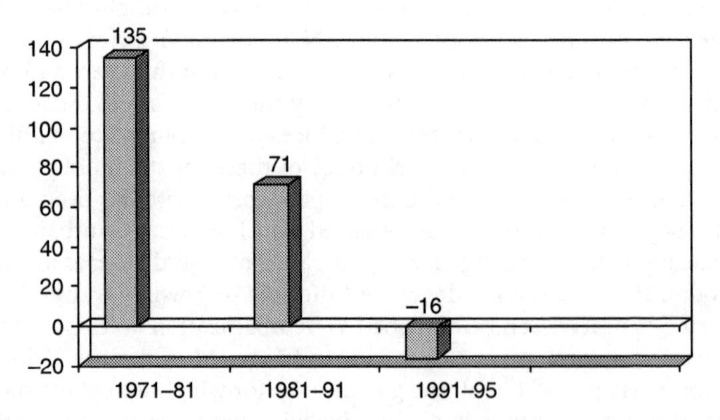

Figure 2.4 Annual increase in per capita car vkt, 1971–95: Melbourne, Perth, Brisbane, Sydney

(Webber, 1968). More recently the contradictory aspects of information technology have been recognized. IT has the ability to reform urban economies based on the simultaneous power to reduce some face-to-face interchanges (routine and follow-up communications) while generating a (perhaps greater) need for face-to-face interchanges where creative interaction and sharing of skills is required (Castells, 1989; Castells and Hall, 1994). Girouard (1985: 381) declares, 'No form of technology is ever going to get rid of the value of face to face contacts in government, education or business, or of the tendency of people of a certain kind to congregate where the options are richest; and as long as this is the case cities of some form will continue.' Hall (1997), after several years of being very equivocal on this,

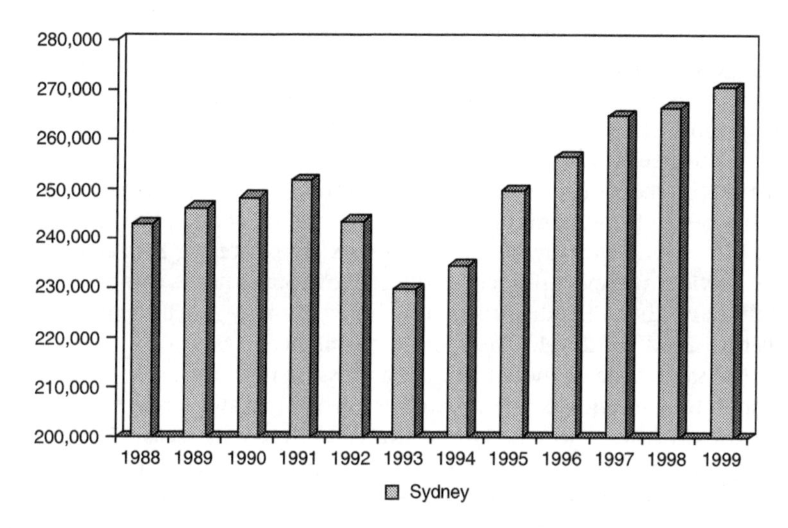

Figure 2.5 Trends in rail use in Sydney, 1988–2000

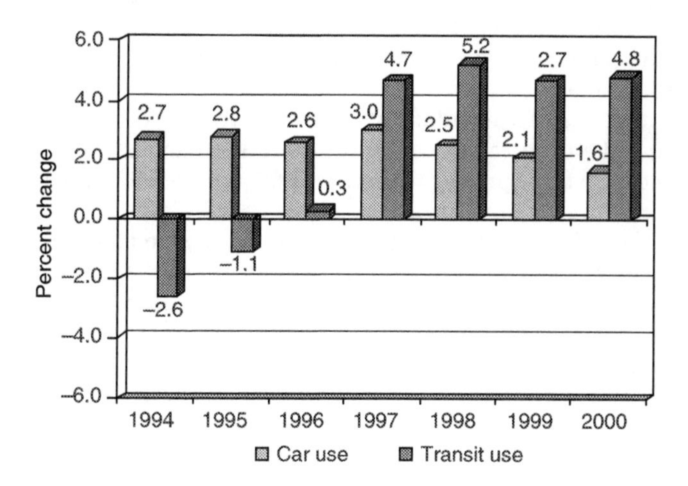

Figure 2.6 US trends in car and public transport use
Source: Smart Growth America.

now states: 'The new world will largely depend, as the old world did, on human creativity; and creativity flourishes where people come together face-to-face' (p. 89). Evidence from our data collection shows that most cities that are part of the global information age are re-concentrating around urban centres (Newman and Kenworthy, 1999). The more global the city the more it is concentrating into these nodes.

The forces that appear to be behind this are as old as the city itself. Although we have the technological means to interact through computers and telephones, we still need personal contact. Thus businesses which are part of the new global information age need to meet others that complement their skills. For example, architectural firms, engineering firms, graphic design firms, computer firms, all need to meet and plan business. So also, professions with overlapping interests are clustering into nodes, and other services and even residences then follow. Some nodes can be based on a dominant kind of industry such as biotechnology – Willoughby (1994) calls this trend the development of 'local milieux'.

At the same time as there has been a pressing need for face-to-face interaction in the new global cities, there has been a shift away from smoke-stack industries in cities, particularly in central and inner city areas. The shift has meant that it is much more attractive for people who need to meet regularly in central or inner areas to locate their housing there as well as their businesses. Thus nodes of information-oriented work mixed with housing and recreation services are becoming an important feature of cities in the last part of the 20th century (Winger, 1997; Newman *et al.*, 1997).

These nodes can form in the inner city or in the suburbs; in the US they tend to be forming in 'edge cities' on the periphery. Data shows that this concentrating force is now of considerable significance in all global cities. It is the process that has revived the inner city in Australia (Sydney has now returned to the density it had in the 1960's mostly due to reurbanization of its central and inner area). Once this was considered impossible for US cities but it appears to be beginning there as well (Norquist, 1998). It has been an on-going process in European cities.

Thus it is possible to see that there is a growing trend in planning for cities in the information age towards the development of urban centres for the face-to-face, creative development of knowledge-based services. This has been called the 'village-isation' of cities. Its support in the movement called 'new urbanism' in the US has reached evangelical proportions. It is essentially designing cities to be less car-dependent. However the evidence on whether people actually *use* cars less in these new developments, especially the new places like Celebration, Seaside and Kentlands which are transit-oriented but without transit, remains to be seen. Reurbanization with a new urbanist flavour, as has occurred in Australia, is much more likely to be less car-dependent in theory and practice. Again, therefore, there is a social and political overlay to this analysis which cannot be neglected. Cities are still made up of discerning people and if such people are placed in physical proximity, that ought to mean a more urban and less car-oriented lifestyle.

A further set of economic factors are only just emerging. Their role will be uncertain but they have major potential to shape cities.

Oil prices and global warming

If oil prices continue to rise there will be a major economic fallout, especially in the most car-dependent cities and, within them, for the most car-dependent citizens and suburbs. The first signs that this is influencing policy can be seen (see Chapter 3 this volume) and certainly there is strong evidence that whenever an oil crisis has occurred there has been a rapid move upwards in the value of real estate in non-car-dependent areas.

Global warming can impinge on cities directly through increased flooding, even to the extent of large scale immigration to less vulnerable cities, but more likely in the extra costs of rebuilding the city's infrastructure. This longer term trend can be seen also against the likelihood of carbon taxes which will do the same as oil price increases. Both factors would reduce car-dependence. However, whether people in cities see these as likely and even – in the middle of such trends – whether they are prepared to change, is still a serious question. However the evidence in some surveys given below would suggest that a more serious evaluation of car-dependence is in fact occurring.

Governance and urban growth

Governments and town planners are rarely as important in the shaping of cities as they think they are. Grand plans are rarely implemented, though they can be influential, especially if they correspond with the social values of a city. Few cities today have plans which suggest they are wanting to sprawl, with greater car-dependence. Most have plans to develop corridors, subcentres and options that can create more choice in terms of mode of transport. There are as well a number of global governance trends that are influencing this process.

There is a world wide trend towards decentralization of government which parallels the globalization of the economy. This is driven by the need for local economic responses to global trends and the human need to belong to community as national boundaries dissolve. The process of decentralization is associated with 'Third Way' politics and is being driven by legislation like ISTEA and TEA-21 in the USA which empowers communities to have a bigger say in their transport infrastructure (see Chapter 6 this volume). In third world cities it is giving a greater role to local government and less to large national projects like toll roads. This process provides a stronger political base for the social values associated with reduced car-dependence though it can also be used to preserve areas in car-dependence.

Global processes for reducing transport energy are part of the international sustainability agenda. Although largely based on moral persuasion, this agenda can influence trends in cities through priorities in government funding. The Australian government funds around 50 cities across Australia in the ICLEI program, Cities for Climate Protection. However the driving force for urban road building is still unrelated to such an agenda (see Chapter 12 this volume).

Within the transport and planning profession there is an international agenda to try to integrate how cities are planned. There is an awareness of how cities were sprawled unsustainably by large capacity roads and by scattered housing with few services. The need for closely integrated approaches has been highlighted by the New Urbanism in the USA, the 'Better Cities'

Table 2.4 'Smart Growth' values in the United States, 2000

Question	**% of sample**	**% of sample**	**% of sample don't know or non-response**
Do you agree more with those who say it is better to have land-use planning to guide the place and size of development, or more with those who say that people and industry should be allowed to build wherever they want?	Land use planing 78	Allow to build 17	5
Thinking about your state, is there a need to do more or to do less to manage and plan for new growth and development in your state?	More 76	Less 13	11
Should state government give funding priority to maintain services, such as schools and roads, in existing communities, rather than encourage new development in the countryside?	Yes 81	No 14	5
Should there be zones for greenspace, farming or forest outside of existing cities and suburbs that would be off limits for developers?	Yes 83	No 14	3
Should government use tax dollars to buy land for more parks and open space and to protect wildlife?	Yes 77	No 21	2
Should your state government use more of its transportation budget for improvements in public transportation such as trains, buses and light rail, even if this means less money to build new highways?	Yes 60	No 36	4
Should part of the transportation budget be used to create more sidewalks and stop signs in communities to make it safer and easier for children to walk to school, even if this means less money to build new highways?	Yes 77	No 19	4

Source: http://smartgrowthamerica.org/, US National Survey on Growth and Land Development, Sample: 1,007 people 18 years of age and above.

program in Australia, and by a range of integrated programs like the Dutch ABC scheme which has also been adopted in the UK.

In a country, the USA, where planning is seen to be a left-wing conspiracy by many there has been a reaction towards an increased role for governments in city planning. The 'Smart Growth' movement has become very active in ballot initiatives to prevent suburban developments on the urban fringe and to build transit options instead of planned highways (see below). Similar movements are evident in every city around the world, even in third world cities where the SUSTRANS network highlights such issues (http://www.sustrans.org.uk/webcode/home.asp). Whether they are successful depends on their creativity, persistence and the political responses to these social movements.

Changing urban values

Having provided something of the rationale for why people may be changing in their urban and transport values, it is possible now to look at some of the patterns that are emerging in the stated values of city dwellers around the world.

The changing nature of urban values is part of the global trend to ecosocialization of values that embrace greater diversity and greater choice. It is being expressed in all kinds of ways but they are mostly more urban than suburban. In particular they are being expressed in movements that are opposed to the simplistic modernism of freeway construction, and thus they have come hard up against the potential growth in mobility.

Another part of this new lifestyle-oriented movement is the importance of health. The importance of walking in a person's daily life, the need for a less polluted and less stressed urban environment, are highly valued aspects of lifestyle. The new element is the recognition that the suburbs may not be the best expression of this need, and that certainly the continued provision of car-based infrastructure and endless sprawl, are not helping in this quest. What follows is a selection of the kinds of attitudes being found in cities with reference to issues of 'Smart Growth' and transport.

'Smart Growth' transport values

In the USA a survey was conducted by the Smart Growth America Coalition in 2000 (see website <http://smartgrowthamerica.org/>). Answers to the question: 'which one of the following proposals is the best solution for reducing traffic congestion in your state?' were as follows: 'build new roads': 21 per cent, 'improve transit': 47 per cent, 'develop communities': 28 per cent. Further results are set out above in Table 2.4. The survey shows that the public clearly want to see intervention to prevent cities sprawling further, they want to see a change in transport infrastructure priorities towards more public transport, bicycling and walking, and they are happy to see

Table 2.5 Department of transport Perth telephone survey of 400 people

Transport priority	% response
Improve buses	55
Improve trains	40
Improve existing road conditions	30
Improve and construct footpaths	19
Traffic calming and management	18
Build new roads/extend roads	9

Source: Western Australia Department of Transport.

these funds come out of traditional road fund sources (see also Chapter 6 this volume).

The trend in US politics has been to try and empower local communities in their transport infrastructure decisions. However if the trend in government continues to favour road lobbies, or if the social problems of the inner city are not solved to enable reurbanization to occur, then the outcome could be greater car-dependence. To make an impact social values have to translate into political reality.

A similar survey in Perth on transport priorities found similar responses but even more strongly in the anti-car-dependence direction (Table 2.5).

Surveys in Europe (e.g. Gotz *et al.*, 1997; Brög, 2001) continue to show a much higher commitment to public transport and walking/cycling than in the New World. Car ownership has continued to grow in the past few decades and along with it car use, but there are also many cities where this growth has stopped or reversed, including wealthy cities like Zürich, Copenhagen and Stockholm (Newman and Kenworthy, 1999). The provision of good transit options has been successful in almost every case. In the UK where little of innovative transit technology has been attempted, Klau (2000) has found in surveys of three cities that 'there is a substantial latent demand for public transport, especially for a frequent, reliable, and easily reachable service. Light rail has a strong potential following among car users, even in cities with no recent experience of light rail or tram (p. 120).'

The new phenomenon in European urban transport values and planning is the 'car-free housing' movement. Examples of 'car-free' housing are appearing all through Europe with various ways of building without providing car parking (thus freeing up around one third of the site's space). The use of car-sharing clubs and transit passes paid for from leases or rents is one of the key approaches. When surveyed it was found that 45 per cent of the car-free households had sufficient income to purchase a car but preferred not to. These households are more able to live a car-free lifestyle in European cities but the trend to try and emulate this in American and Australian cities has begun, based around car-sharing arrangements (Scheurer, 2001).

Conclusions

A simple projection of past mobility trends would suggest that car use in cities is likely to continue to grow rapidly. Conventional wisdom suggests that mobility is linked directly to wealth. However, it is possible to see a change in values, towards ecosocial values. Some emerging trends would suggest mobility in cities may be uncoupled from wealth. A similar pattern has occurred over energy use. Simple projections in the 1960s predicted massive growth. However even the 'soft energy path' prophets of that period like Amory Lovins were not able to predict the extent of the shift towards greater efficiency and distributed energy systems that has occurred. Energy use became essentially decoupled from wealth.

This chapter has suggested that mobility may now be being decoupled from wealth, at least in cities. The city of the future appears to be becoming more urban, less car-dependent, more transit-based, more oriented to local centres of knowledge-based employment and community activity. This is being driven by ecosocial values linked to limits being reached by transport technology, ecosocial values in the new knowledge economy and ecosocial values associated with urban lifestyles. Understanding the dynamics of cities, especially the dynamics of how time budgets are valued, enables us to see that the constant spread of cities and the loss of options in fast, reliable transit, will not be socially and politically acceptable for long. Reurbanization and reorientation of transport priorities are thus happening in most cities around the world. Both are understandable as responses to car-dependence or the increase of *urban* rather than suburban values.

Such a future is not to be feared as it is far more sustainable. But it is also not assured. Ecosocial values can shift quickly into anti-urbanism and fear with a consequent rapid increase in car-dependence as people flee dysfunctional cities. This will happen if governments, communities and the private sector cannot grasp the new ecosocial urban agenda and respond to the values being expressed in this chapter. Dysfunctional cities based on fear, litter human history.

Note

1. The books we have written are based on data linking car use to urban density, centrality, mixed use and transit availability (Newman and Kenworthy, 1989, 1999; Kenworthy *et al.*, 1999). A recent survey of ours has found that 80 per cent of the variation in transport patterns in Australian cities can be explained by these urban form variables.

3
The Peak of Oil: an Economic and Political Turning Point for the World

C.J. Campbell

Introduction

Concerns about global warming and its consequences have received much worldwide attention, but little consideration has been given until recently to the issue of oil depletion and the consequential peak of production, which may have a more immediate and greater impact (Youngquist, 1997; Campbell and Laherrère, 1998).

Oil fuelled the economic prosperity of the 20th century, but production is now close to peak and cannot see us through the 21st century. Prices rose markedly in 2000 prompting a fall in demand. More price shocks are in store, and chronic shortage will appear before long. Agriculture has been described as a process that converts petroleum into food, insofar as it relies heavily on petroleum for synthetic nutrients, mechanised farming, transport and storage, as well as, indirectly, the supply of water. The new crops of the 'green revolution' have a voracious appetite for nutrients and water, so our very food supply depends on petroleum.

The world is run on the classical economic principles of supply and demand as if everything were just a matter of human intervention and transaction. According to this way of thinking, if wheat prices rise, the farmer sows more seeds in a self-adjusting mechanism. But oil is extracted not produced, as it is a finite resource. This challenges economic theory, but its adherents counter by claiming that extraction is a matter of concentration, as is the case for coal or copper, so that if prices rise or costs fall, lower concentrations of ore become viable. This happy solution however fails because oil is a liquid not an ore. The oil–water contact at the base of the oil accumulation in a field is abrupt, so there is no possibility of tapping lower concentrations. We need to understand the occurrence of oil in nature.

Geological origins of petroleum

We may start the discussion with a brief summary of the geology of petroleum, by which we mean everything from sticky tar through normal crude oil, to gas and the liquids that may be recovered from gas. A geochemical breakthrough in the 1980s greatly clarified our knowledge of the origins of petroleum that previously had been rather hazy (Campbell, 1997). Isotopic analysis made it possible to relate the oil in a well to the source-rock from which it came; and laboratory experiments provided knowledge of the chemical reactions that converted the organic material into oil and gas. Lastly, advances in geology gave insight into the conditions under which these very rare hydrocarbon source-rocks were deposited.

Briefly, oil comes from algal material, whereas gas comes from plant remains and deeply buried oil where high temperature breaks the molecular bonds. It transpires that the great bulk of the world's oil comes from only a small number of periods of extreme global warming, when seas and lakes became poisoned by excessive algal blooms. The soft remains of these organisms sank to the floors of the seas and lakes in which they lived. In many cases, they were oxidised by currents or consumed by bottom-living organisms, but where they fell into stagnant troughs, commonly formed by rifts as the geological plates moved apart, they were preserved and later buried beneath younger sediments. The two last epochs of prolific generation occurred about 90 million years ago, giving the huge deposits of the Caribbean borderlands, and 145 million years ago, giving those of the Middle East, North Sea and much of Russia.

As the rifts continued to subside, they were filled by an overburden of sediment, washed in from the surrounding lands, and the organic material was subjected to the earth's heat-flow, which initiated chemical reactions that converted it into oil and gas. Critical exposure to heat, from a combination of high temperature and long exposure to it, gave rise to the so-called 'oil window' when prolific amounts of oil were formed. Generally, it occurs when the source-rocks have been buried to depths of 2,000 to 5,000 metres, depending on the geothermal gradient. Gas, being derived from plant remains, was more widely generated than was oil.

Generation occurs under very high pressures forcing the oil and gas to begin to migrate upwards through the rocks along hair-line fractures. In some cases, the charge reaches equilibrium to form a dissipated deposit, but in other cases it encounters a porous and permeable stratum that provides a conduit through which it can flow. The pore-space in the conduit is filled with water, left over from the sea or lake in which it was deposited, and the oil and gas, being lighter, move upwards. If the conduit leads to the surface, the oil and gas are lost to the atmosphere. But if, as is commonly the case, it has been affected by subsequent earth movements, the oil and gas will collect at the top of folds, up against faults or where the conduit itself dies out.

These traps, which form oil fields, also rely on being capped by a seal, of which the most effective is salt, although leakage over geological time is inevitable, with gas being more vulnerable than oil. Furthermore, traps once formed may be subjected to later earth movements causing the hydrocarbons to re-migrate, either to escape or collect in other traps.

This brief and simplified account is enough to explain why only a small fraction of the hydrocarbons generated in nature are preserved in traps large enough to form commercial deposits, and why only traps filled in the comparatively recent geological past still hold their charge. It also readily explains why oilfields are clustered together in areas having the right geology, separated one from another by vast barren tracts.

We should also mention – and confidently dismiss – a rival theory advocated by Thomas Gold, a provocative academic, that hydrocarbons are of primordial origin emanating from deep in the Earth crust (Gold, 1999). While running in the face of the oil industry's practical experience, this theory is sometimes paraded by 'flat-earth' economists desiring support for the notion of infinite resources that they need to underpin their classical economic theories (Odell, 1992).

In earlier years, knowledge of oil geology was based primarily on field surveys of outcrops at the surface, which indeed found most of the world's oil provinces. Later came geophysical surveys, making it possible to map the subsurface in extreme detail, aided by colossal computing power. We can see the needle in a haystack, but ironically, these great achievements have somewhat diminished the perceived scope for discovery by not only showing the prospective areas but also confidently ruling out the non-prospective barren tracts. The needle is still a needle however well you can see it. We did not need this resolution in earlier years when most of the world's giant fields, holding most of its oil, were found.

Two simple questions: how much has been found? And when was it found?

Having now described, albeit in the simplest terms, the occurrence of hydrocarbons in nature, it is time to move on to ask two very elementary questions: How much has been found? And when was it found?

But before we can attempt to answer these questions, we need to ask what exactly to measure. It is a question that every butcher asks when he decides whether to weigh just the meat or to include the bones. There is some nourishment in bones but a juicy steak is easier to digest. So it is with oil. There are many different categories, each with its own endowment in nature, its own costs and characteristics, and above all, its own depletion profile. Each category is itself finite, and the extraction rates differ. An uncontrolled gas well depletes itself rapidly in one great puff, whereas the production of a tar deposit lasts for a long time, rising to no more than a low peak. A moment's reflection tells us that production of any finite resource starts and ends at zero, reaching a peak in between more or less at the half way mark, called the midpoint of depletion, when past production equals what is left for the future. We are not so much interested in when the resource will be finally exhausted which is a somewhat academic concept, but we should be most concerned to know when production will reach a peak. For reasons to be discussed later, the peak will be a turning point for the world of much greater impact than producing the last barrel.

We return now to the issue of measurement to decide what is figuratively meat and what are bones. It is common practice to distinguish *conventional* from *non-conventional* oil and gas. Approximately 95 per cent of all oil produced to-date can be classed generally as *conventional*. It will continue to dominate all supply for many years to come and is what matters most in determining peak, but failure to define exactly what is meant by the term is a source of great confusion and misunderstanding. Here, we exclude the following categories to be classed as *non-conventional*:

- oil from coal and 'shale' (actually immature source rock);
- bitumen and extra-heavy oil (<10° API);
- heavy oil (10–17.5° API);
- polar oil and gas;
- deepwater oil and gas (>500 metres water depth);
- various types of gas such as coal-bed methane, gas in tight reservoirs, gas in geo-pressured aquifers, gas hydrates.

(Degrees API, standing for American Petroleum Institute, is an oil-industry term used as a measure of density.)

A further complexity arises from the liquids that condense from gas on being brought to the surface or are extracted by processing. They belong to the gas domain, but are often metered with oil.

These *non-conventional* classes will be discussed in a later section. Here, we address our primary concern, namely *conventional* oil, and the question of how much has been found in the world's 20,000 oilfields. Technical estimates of the size of a field are relatively constant although they naturally evolve as knowledge grows, but the reporting of reserves is an entirely different matter being much influenced by commercial, political and regulatory factors (Laherrère, 1999).

The starting point is a geological prospect, whose size is determined from geophysical and geological mapping with a range of assumptions concerning the reservoir and recovery factor. This estimate, which can be described as a scientific estimate, is confidential to the oil company, and does not accordingly enter the public database.

If the first well on the prospect, termed a '*wildcat*', makes a discovery, the emphasis shifts from determining the ultimate size of the field to designing an initial development programme to reduce risk and increase economic

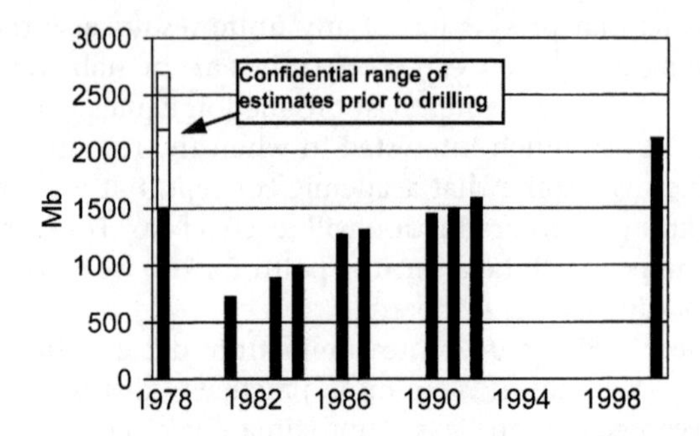

Figure 3.1 The Oseberg field in Norway: the practice of reserves reporting. The early confidential range of estimates of the size of the field were accurate. The reserves additions of each successive phase of development were reported giving the misleading impression of growth as if it were a technological dynamic to be projected forward. In fact, the field ends up holding about as much as originally estimated

return. The plans involve setting the optimal plateau of production for which to design the facilities, as it clearly makes no sense to invest for a short-lived peak.

It is normal for the company to announce the reserves attributable to the plan for the first phase of development, which duly find their way into the national statistics. Both the plan and the development itself are, however, subject to progressive modification as production proceeds and as efforts shift to prolonging the plateau for as long as possible, which makes good economic sense. This may be achieved in a number of ways, including building new platforms, drilling wells with subsea completions, and by tapping subsidiary reservoirs and traps, both within the field and in its immediate vicinity (see Figure 3.1).

As this work proceeds, the reported reserve estimates are revised upwards giving the impression of growth, which is commonly attributed to technology and managerial skill. Furthermore, companies have every incentive to under-report the initial size of their reserves and see them grow by revision because that delivers a much more favourable image to the investment community than one champagne party, followed by a sad decline. There are tax advantages too in countries having a depletion allowance based on reported reserves.

Understanding the practice of reserve reporting is of cardinal importance. It is one of the main factors to have misled the 'flat-earth' economists, who attribute growth to technology which they believe can be extrapolated ever onward. In reality, an oil field contains what it contains because it was filled in the geological past. Extending plateau production yields more profit but accelerates depletion. The impact on the reserves themselves is negligible.

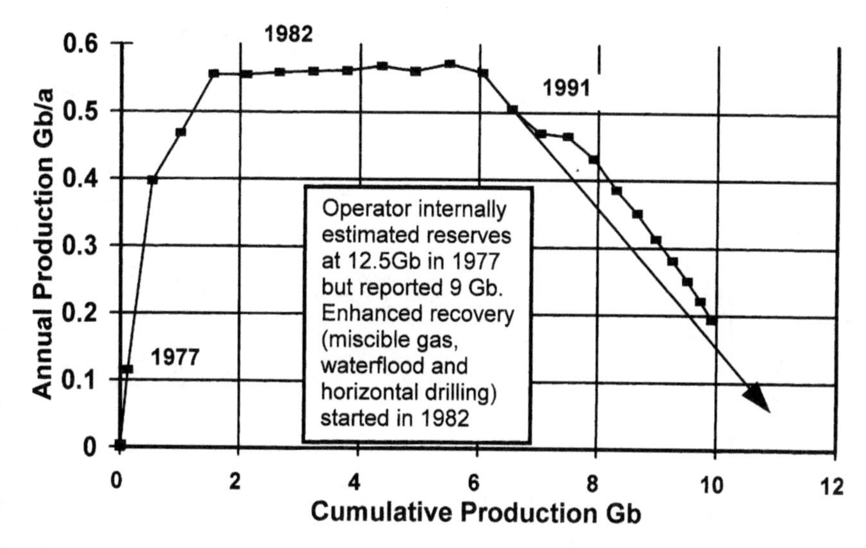

Figure 3.2 The Prudhoe Bay field in Alaska, demonstrating that technology has had no impact on reserves

In short, the industry has treated reserves from discovery as a form of inventory to be drawn down as best serves financial purposes. No conspiracy is implied: the practice was consistent with prudent management; it better complied with stockmarket regulations designed to prevent fraud; it gave a better stockmarket image; it reduced tax and it made the company generally less vulnerable to hostile takeover or expropriation. But for the purpose of estimating future production, we need to know the correct reserve numbers and the correct dates of discovery.

The point is well illustrated in Figure 3.2, which shows annual production against cumulative production for the Prudhoe Bay Field in Alaska. In 1977, the operating company internally estimated its reserves at 12.5 Gb (gigabarrels – billion barrels) but prudently reported 9 Gb. Enhanced recovery technology, involving gas injection and later horizontal drilling, commenced in 1982, but plateau production continued until 1989, when decline set in. The enhanced recovery succeeded in arresting decline for one year, but the subsequent decline became steeper. It is now evident that the field will barely make the original estimate, unequivocally showing that technology has had a negligible impact on reserves. This experience is typical of many large fields.

For all these reasons, the reported reserves of a field have generally grown over time. The key word is *reported,* as nothing was added in nature. A moment's reflection confirms that a field contains what it contains because it was filled in the geological past, even if that is not necessarily known or reported initially. A second moment of reflection tells us that the revisions

Table 3.1 Conventional oil endowment*

	Summary				Conventional oil endowment Base case scenario						Revised		14 August 2001			2000
	Country	Production Cum.			Reserves: Reported					Discovered	Yet-to-find	Yet-to-produce	Ultimate	Dep. rate (%)	Dates	
		kb/d 2000	Prod. Gb	5 yr Trend (%)	World oil	O&GJ	Adjust +/−	Factor	Assessed reserves						MP dep.	Peak prod.
1	Saudi Arabia	8,064	**88.9**	1	259.10	259.20	0.00	0.80	**207.4**	296.3	**23.7**	231.1	**320**	1.3	2016	**2010**
2	Russia	6,351	**119.0**	1	52.66	48.57	−2.32	1.2	**55.5**	174.5	**15.5**	71.0	**190**	3.2	1990	**1987**
3	USA-48	4,096	**167.1**	−4	19.63	21.76	−9.00	1.5	**19.1**	186.2	**3.8**	22.9	**190**	6.1	1970	**1971**
4	Iraq	2,681	**25.9**	73	100.00	112.50	−2.67	0.85	**93.4**	119.2	**15.8**	67.5	**135**	0.9	2022	**2010**
5	Iran	3,568	**51.7**	−1	93.10	89.70	−2.58	0.75	**65.3**	117.1	**7.9**	80.8	**125**	1.7	2007	**1974**
6	Venezuela	2,580	**44.5**	1	47.1	76.9	−30.00	0.9	**42.2**	86.7	**8.3**	50.5	**95**	1.8	2003	**1968**
7	Kuwait	1,774	**29.6**	−0	92.40	94.00	−5.49	0.55	**48.7**	78.3	**6.7**	55.4	**85**	1.2	2012	**2011**
8	Abu Dhabi	1,850	**16.7**	−1	62.50	92.20	−6.77	0.65	**55.5**	72.2	**4.8**	60.3	**77**	1.1	2020	**2011**
9	Mexico	3,050	**27.7**	1	28.3	28.3	0.00	1.0	**26.8**	54.5	**2.5**	29.3	**57**	3.7	2001	**2001**
10	China	3,255	**26.1**	1	34.1	24.0	−12.15	2.0	**23.7**	49.8	**5.2**	28.9	**55**	4.0	2001	**2001**
11	Libya	1,408	**21.9**	0	29.5	29.5	−1.51	0.9	**23.8**	45.7	2.3	26.1	**48**	1.9	2004	**1978**
12	Nigeria	1,991	**20.9**	−2	24.5	22.5	−2.44	1.3	**25.1**	46.0	**2.0**	27.1	**48**	2.6	2004	**2004**
13	Kazakhstan	627	**5.4**	7	6.42	5.42	−0.23	5.0	**10.4**	15.8	**24.2**	34.6	**40**	0.7	2027	**2032**
14	Norway	3,216	**14.0**	1	10.0	9.4	0.00	1.5	**14.2**	28.1	**1.9**	16.0	**30**	6.8	2001	**2001**
15	UK	2,537	**18.0**	−0	5.0	5.0	0.00	2.0	**10.0**	28.0	**1.0**	11.0	**29**	7.7	1996	**1999**
16	Indonesia	1,299	**19.0**	−1	8.4	5.0	−2.97	4.0	**8.02**	27.0	**1.0**	9.0	**28**	5.0	1990	**1977**
17	Canada	1,118	**18.0**	−7	5.58	4.71	0.00	1.8	**8.47**	26.5	**1.5**	10.0	**28**	6.1	1990	**1973**
18	Algeria	800	**11.5**	−0	13.0	9.2	−2.58	2.0	**13.2**	24.8	**1.7**	15.0	**26.50**	1.9	2005	**1978**
19	Azerbaijan	258	**7.9**	8	0.00	1.18	−0.09	5.0	**5.42**	13.3	**1.69**	7.11	**15.00**	1.3	1995	**2005**
20	Qatar	681	**6.3**	8	5.4	13.2	0.00	0.7	**9.21**	15.5	**0.5**	9.7	**16.00**	2.5	2006	**2006**
21	N.Zone	628	**6.2**	6	4.65	5.00	−1.61	2.00	**6.77**	13.00	**1.00**	7.78	**14.00**	2.9	2003	**2003**
22	Oman	891	**6.3**	0	5.7	5.5	0.00	1.2	**6.61**	12.9	**0.9**	7.5	**13.75**	4.2	2002	**2002**
23	Egypt	811	**8.1**	−1	3.8	2.9	−0.30	1.5	**3.98**	12.1	**0.6**	4.6	**12.75**	6.0	1994	**1994**
24	Argentina	749	**7.7**	−1	2.6	3.0	0.00	1.0	**2.89**	10.6	**0.4**	3.3	**11.00**	7.7	1992	**1998**

25	India	639	5.1	−1	3.4	4.7	0.00	1.1	4.96	10.1	0.7	5.6	10.75	4.0	2001	2001
26	Colombia	689	5.3	2	2.3	2.6	−0.55	2.0	4.06	9.4	1.1	5.2	10.50	4.6	2000	1999
27	Angola	744	4.0	2	8.5	5.4	−4.21	5.0	6.01	10.0	0.5	6.5	10.50	4.0	2003	2003
28	Australia	720	5.3	7	2.9	2.9	−0.46	1.2	1.55	8.25	0.50	3.43	8.75	7.1	1995	2000
29	Malaysia	671	4.8	1	4.6	3.9	−0.77	0.9	2.82	7.57	0.43	3.25	8.00	7.0	1997	1999
30	Brasil	413	4.3	4	8.1	8.1	−9.00	−3.0	2.70	6.97	0.53	3.23	7.50	4.4	1997	1985
31	Romania	123	5.7	−2	1.2	1.4	−0.09	0.9	1.20	6.87	0.63	1.84	7.50	2.4	1973	1976
32	Ecuador	390	3.0	0	3.0	2.1	−0.70	3.0	4.25	7.24	0.26	4.51	7.50	3.1	2005	2005
33	Turkmenistan	143	2.8	15	–	0.5	−0.05	4.0	1.98	4.79	1.21	3.18	6.00	1.6	2003	1973
34	Syria	510	3.4	−2	2.3	2.5	−1.23	1.2	1.53	4.94	0.56	2.09	5.50	8.2	1997	1995
35	Dubai	333	3.6	5	1.0	4.0	−1.53	0.4	0.99	4.55	0.20	7.46	4.75	9.3	1990	1991
36	Brunei	169	2.9	2	1.0	1.4	−0.58	2.0	1.55	4.41	0.09	1.64	4.50	3.6	1987	1978
37	Gabon	331	2.6	−2	2.6	2.5	−1.37	1.2	1.35	4.00	0.25	1.60	4.25	7.0	1996	1996
38	Peru	97	2.3	−4	4.1	0.3	0.00	5.0	1.55	3.81	0.19	1.74	4.00	2.0	1994	1983
39	Ukraine	48	2.6	−9	–	0.4	0.02	1.2	0.49	3.09	0.91	1.40	4.00	1.2	1983	1970
40	Trinidad	120	3.1	−1	0.7	0.7	0.00	1.0	0.69	3.80	0.20	0.89	4.00	4.7	1980	1978
41	Yemen	441	1.4	4	2.1	4.0	−1.25	0.5	1.37	2.77	0.48	1.86	3.25	8.0	2001	2001
42	Uzbekistan	152	0.9	−1	–	0.6	−0.06	2.5	1.35	2.29	0.71	2.05	3.00	2.6	2008	2008
43	Congo	265	1.3	4	1.7	1.5	−0.51	1.5	1.49	2.82	0.18	1.67	3.00	5.5	2002	2002
44	Vietnam	304	0.7	15	1.8	0.6	−0.29	5.0	1.56	2.22	0.53	2.09	2.75	5.0	2005	2005
45	Denmark	357	1.1	14	0.9	1.1	−0.13	1.1	1.03	2.10	0.40	1.43	2.50	8.3	2001	2001
46	Germany	61	1.9	1	0.3	0.4	0.00	1.0	0.38	2.26	0.14	0.52	2.40	4.1	1976	1967
47	Tunisia	75	1.1	−3	0.3	0.3	−0.06	2.6	0.65	1.80	0.20	0.86	2.00	3.1	1995	1995
48	Italy	92	0.8	−3	0.6	0.6	−0.07	1.8	0.99	1.81	0.19	1.18	2.00	2.8	2005	2005
49	Thailand	110	0.3	16	0.3	0.4	0.00	2.5	0.88	1.21	0.29	1.17	1.50	3.3	2008	2008
50	Cameroon	101	1.0	−2	0.6	0.4	−0.61	−1.5	0.32	1.31	0.19	0.51	1.50	6.7	1994	1986
51	Bahrain	102	1.2	−0	–	0.1	−0.04	1.0	0.11	1.27	0.23	0.34	1.50	9.8	1985	1993
52	Turkey	57	0.8	−3	–	0.3	0.00	0.8	0.24	1.03	0.17	0.41	1.20	4.8	1992	1991
53	Hungary	27	0.7	−2	0.1	0.1	−0.01	0.9	0.09	0.78	0.22	0.31	1.00	3.1	1986	1994
54	Sharjah	45	0.4	−7	–	1.5	−0.19	0.3	0.39	0.83	0.17	0.56	1.00	2.8	2003	1997
55	Croatia	23	0.5	−5	0.0	0.1	−0.01	4.0	0.33	0.81	0.19	0.52	1.00	1.6	2003	1977
56	Netherlands	51	0.8	−2	0.1	0.1	−0.02	1.4	0.12	0.93	0.07	0.19	1.00	8.8	1987	1989
57	Austria	20	0.8	0	0.1	0.1	−0.01	1.3	0.10	0.87	0.08	0.19	0.95	3.8	1971	1955
58	France	29	0.7	−7	0.2	0.1	0.00	1.2	0.17	0.88	0.07	0.25	0.95	4.1	1987	1988
59	Papua	69	0.3	−8	0.0	0.4	−0.06	1.3	0.39	0.70	0.20	0.60	0.90	4.0	2005	1993

Table 3.1 Continued

	Summary				Conventional oil endowment Base case scenario						Revised			14 August 2001		2000
	Country	Production Cum.			Reserves					Discovered	Yet-to-find	Yet-to-produce	Ultimate		Dates	
					Reported											
		kb/d 2000	Prod. Gb	5 yr Trend (%)	World oil	O&GJ	Adjust +/−	Factor	Assessed reserves					Dep. rate (%)	MP dep.	Peak prod.
60	Bolivia	34	**0.4**	4	0.2	0.4	0.00	0.6	**0.24**	0.64	**0.16**	0.40	**0.80**	3.0	2000	**1974**
61	Albania	6	**0.5**	−7	–	0.2	−0.04	1.2	**0.15**	0.68	**0.12**	0.27	**0.80**	0.8	1986	**1983**
62	Pakistan	55	**0.3**	−1	0.2	0.2	−0.08	1.3	**0.17**	0.59	**0.16**	0.33	**0.75**	5.7	1998	**1992**
63	Philippines	1	**0.0**	−3	0.0	0.3	−0.00	1.2	**0.35**	0.39	**0.21**	0.55	**0.60**	0.1	2013	**2013**
64	Chile	9	**0.4**	−1	0.1	0.2	−0.01	0.3	**0.04**	0.46	**0.04**	0.08	**0.50**	3.7	1979	**1982**
Regions																
1	ME Gulf	18,565	**219.0**	2.7	611.75	652.60	−19.13	0.75	**477.0**	696.1	**59.9**	537.0	**756.0**	1.2	2016	**2011**
2	Eurasia	11,012	**172.1**	1.4	94.46	82.49	−15.02	1.49	**100.6**	272.7	**50.6**	151.2	**323.3**	2.6	1997	**1987**
3	N. America	5,214	**185.1**	−3	25.20	26.47	−9.00	1.6	**27.6**	212.7	**5.3**	32.9	**218.0**	5.5	1972	**1972**
4	L. America	8,131	**96.6**	0.6	96.48	122.50	−40.26	1.04	**85.4**	182.0	**15.8**	101.2	**197.8**	2.8	2001	**1998**
5	Africa	6,524	**72.6**	−0.8	84.37	74.27	−13.60	1.25	**75.9**	148.5	**8.0**	83.9	**156.5**	2.8	2002	**2004**
6	W. Europe	6,363	**38.0**	0.7	17.19	16.86	−0.23	1.62	**27.0**	65.0	**3.8**	30.8	**68.8**	7.0	1998	**2000**
7	Far East	4,035	**38.8**	1.3	22.57	19.66	−5.20	1.63	**23.6**	62.4	**4.1**	27.7	**66.5**	5.0	1997	**2000**
8	ME. Other	3,060	**23.3**	1.7	16.50	31.11	−4.24	0.76	**20.4**	43.8	**3.2**	23.6	**47.0**	4.5	2000	**2000**
9	Other	347	**3.3**	4.6	0.00	0.00	0.00	0.00	**7.7**	10.9	**1.1**	8.7	**12.0**			
	Unforeseen										**4.0**	4.0	**4.0**			
	Non-swing	44,743	**630**	0.3	357	373	−88	1.3	**368**	998	**96**	464	**1094**	3.4	1997	**1998**
World		63,308	**849**	0.9	969	1026	−107	0.9	**845**	**1694**	**156**	**1001**	**1850**	2.3	2003	**2003**

Note

* *ME Gulf* is Abu Dhabi, Iran, Iraq, Kuwait and Saudi Arabia; *Eurasia* is the former communist bloc of Eastern Europe, Russia and China; *N. America* is Canada and the USA; *W. Europe* is Europe from Norway to Italy; *Other* is countries with relatively small current production; *Unforeseen* is to provide for rounding in the model.

have to be backdated to the discovery of the field to obtain a genuine discovery trend. That, however, is easier said than done because national statistics take the revisions when reported, and it is virtually impossible to backdate them without access to the details of the industry database.

In short, *reserves* may variously refer to the best volumetric estimate, based on a scientific evaluation of the relevant parameters in nature, the expected production of planned facilities, the expected results of actual facilities that may be later modified, the latter being duly fine-tuned from experience of actual well performance. Each of these estimates may be entirely valid within its context but does not tell us what we want to know for our purpose. To be quite clear therefore, we here define *reserves* as estimated ultimate recovery, less *cumulative production* to the reference date, such that revisions are statistically neutral. It assumes the application of all appropriate known technology and adequate economic incentives.

Reserves are reported annually by two trade journals, the *Oil and Gas Journal* and *World Oil*, based on compilations of information furnished by governments and industry. The *reserves* are described as *proved* to meet Securities and Exchange Commission (SEC) regulations, but in reality approximate to *proved and probable*, having a probability of occurrence of about 60 per cent.[1]

Table 3.1 gives a listing of assessed reserves by country as herein defined. The reported *proved reserves* are first adjusted to remove the *cumulative production* for any period of unchanged reports, which are clearly implausible, normally

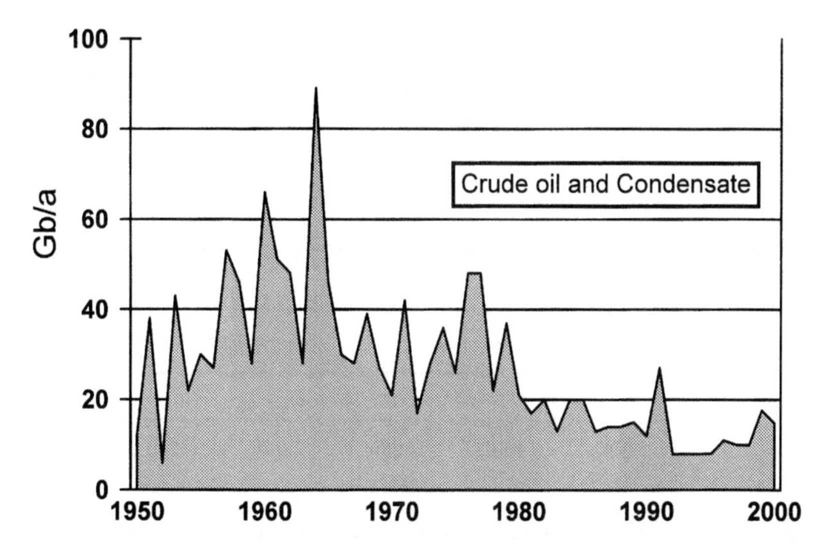

Figure 3.3 The world discovery trend for oil and condensate with revisions duly backdated to the discovery of the fields containing them

reflecting failure by the country concerned to update its estimates. In 2000, as many as 72 countries failed to update their estimates. Then a factor is applied to convert the adjusted value to the best estimate as herein defined.

We can thus answer the two questions posed at the beginning of this section. As of the end of the year 2000, our best estimate is that a total of just over 1600 Gb (billion barrels) of conventional oil had been found since exploration commenced in the middle of the last century, and that discovery peaked in the mid-1960s (Figure 3.3).

How much is yet-to-find?

By yet-to-find we mean the amount of conventional oil to be found and achieve reserve status as defined above. There are several methods for determining it. First, is old-fashioned geological judgment whereby experienced and knowledgeable explorers evaluate a basin on the basis of its underlying geology, the amount of past exploration and the results. They can make valid mental extrapolations and identify untested possibilities. Many reasonably objective studies were made on this basis in earlier years and published (Campbell, 1997). Some early estimates have taken into account simple rock volumes, imagining that a certain volume of rock will deliver

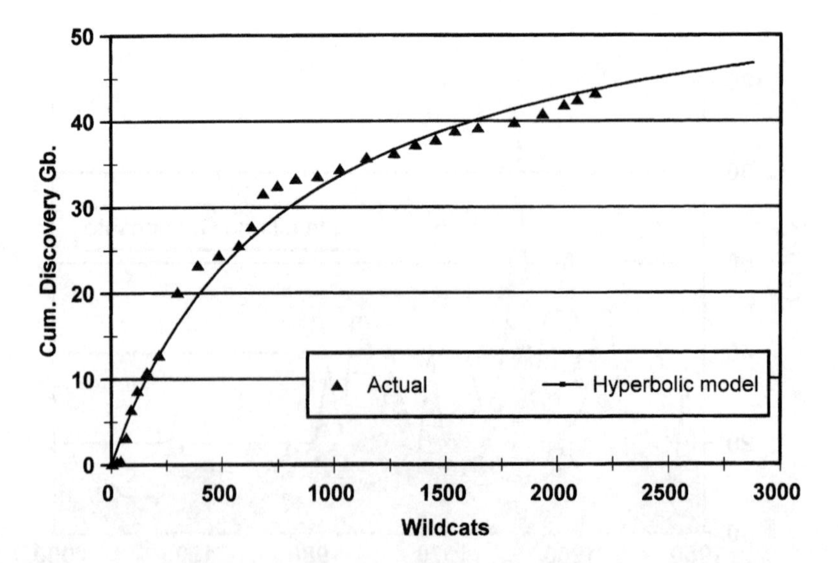

Figure 3.4 The 'creaming curve' of the North Sea plots discovery against wildcats

* The trend is hyperbolic and may be extrapolated to asymptote to show what is left to find, subject to a cut-off when new finds are too small to be economic.

a certain amount of oil, but it is not a good approach because it failed to recognise that the essential source-rocks are so rare and unevenly distributed.

More recently, better statistical methods have been applied, and work well for mature productive basins. Perhaps the most powerful is the so-called *creaming* curve (as in cream off the top of the milk) which plots cumulative discovery against cumulative *wildcats* (being the first borehole on a new prospect) over time. Figure 3.4 shows such a plot for the North Sea, which may be readily extrapolated to show what is left to find in ever smaller fields until viable exploration reaches a limit. Another is the parabolic fractal method that plots size against rank (Laherrère, 1996, 1999). A third relates the cycles of discovery with the corresponding cycles of production.

Estimating the *yet-to-find* is not an exact science, but realistic estimates, as given in Table 3.1, can be made, using knowledge, common sense and the various statistical methods.

The essential parameters

We have now identified the three cardinal parameters describing the endowment in nature. Production eats into *reserves*, and the *yet-to-find* eventually becomes *reserves*, but the total, which is commonly termed the *ultimate*, is a constant.[2] From these parameters, we can derive two further useful elements: how much has been discovered and how much is *yet-to-produce*.

Our best preliminary rounded estimate for end 2000 is as follows (see Table 3.1 for details):

Gb (billion barrels)	
Production in 2000	23
Produced-to-date	850
Reserves	850
Discovered	1,700
Yet-to-find	150
Yet-to-produce	1,000
Ultimate	1,850

Producing what remains

Having now established the amount of *conventional* oil at our disposal, we may turn to consider how it can be produced. If it were contained in a big barrel, someone could push it over and release a flood that would empty it in a flash. If, on the other hand, the barrel were tapped, the flow would be determined by the size of the tap and how far open it was. Emptying the barrel introduces the notion of rate of depletion.

It is broadly the same with the depletion of an oil reservoir far underground, but in this case we have to imagine both that the barrel contains not oil alone but sand impregnated with oil and that the boreholes represent the taps. The oil is now held in the pore space between the sand grains and has to make its way through the myriad of pore-throats to reach the taps. Clearly, the flow-rate is reduced by these constrictions.

Production in an oilfield normally rises as new wells are added until it reaches a peak when the additions from new wells fail to offset the decline from old wells. There is an optimal spacing for the wells: widely spaced wells obviously have longer lives than closely spaced ones with less to drain. In the case of onshore fields close to a market or pipeline, production can commence with the first well and rise to a natural peak before declining, but in an offshore field, production is capped at the designed capacity of the facilities, as discussed above, to give a plateau of production instead of a peak. The production profile of a basin, or a country, is the composite profiles of the individual fields and the rate at which they are found and brought on stream, with the larger fields normally being found first. In a large, single, unfettered environment, the combined production profile from all the fields is a symmetrical bell-curve with peak coming at the mid-point of depletion, when half the *ultimate* has been produced. Individual countries do not always follow such an idealized profile, because there may be more than one cycle of discovery as different geological provinces are opened up, and other factors may intervene. The OPEC (Organisation of Petroleum Exporting Countries) countries have a typical twin-peaked profile with the saddle being due to quota restrictions.

M.K. Hubbert (1956) was one of the first to model production with the help of a bell-curve, when in 1956 he correctly predicted that the Lower-48 States of the USA (mainland states excluding Alaska) would peak around 1970. He was working in a very mature area where discovery had peaked long before in the 1930s, making it relatively easy for him to extrapolate the discovery curve to determine the ultimate. He was able to line up the corresponding production after a time-shift, recognizing that peak would come at the midpoint of depletion when half the ultimate had been produced as illustrated in Figures 3.5 and 3.6.

The difficulty is not so much in the modelling as in the unreliable nature of the data on past discovery, as already discussed. The industry has systematically under-reported the size of discovery for good commercial and regulatory reasons and several OPEC countries have exaggerated their reserves as they vied with each other for production quota which was partly based on reserves. We need the skills of the detective more than the scientist or engineer to unravel all this.

The illusion of technology

No one disputes the enormous technological advances of the oil industry over the past half-century. One radical breakthrough was the development of the

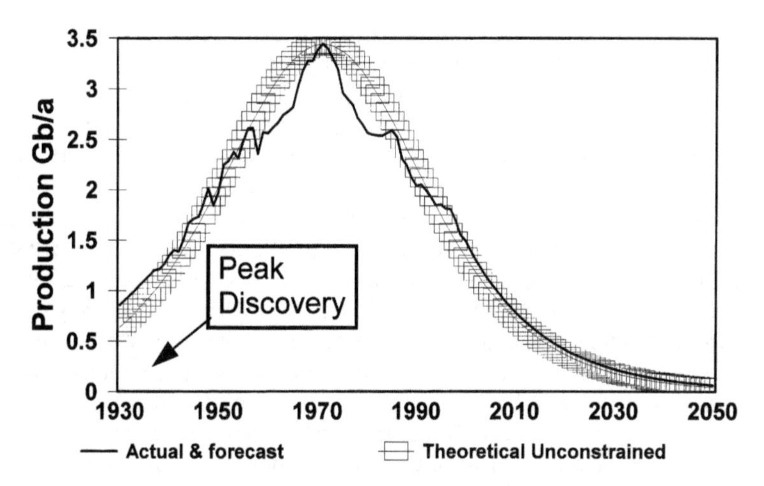

Figure 3.5 Production in the US lower-48 states with superimposed theoretical bell curve for unconstrained production. Discovery peaked around 1930 and production around 1970. If more could have been found, it would have been.

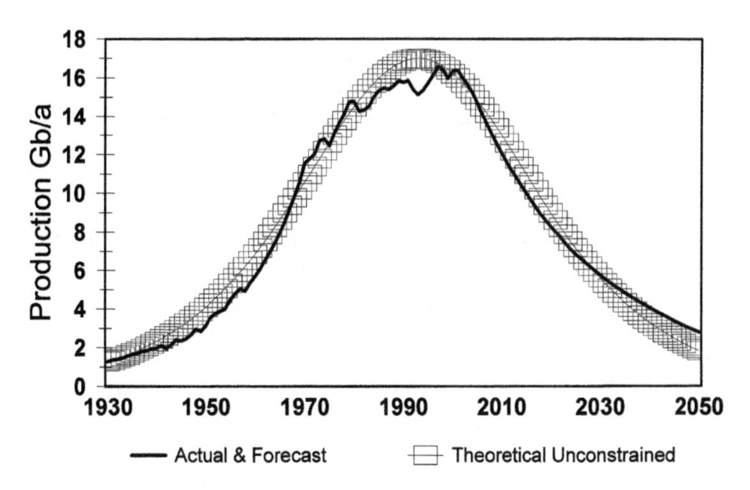

Figure 3.6 Non-swing production with superimposed theoretical bell curve for unconstrained production

semi-submersible rig, whereby a rig was mounted on submerged pontoons beneath the wave base. It opened the offshore to routine exploration. But for the most part the developments were refinements of existing methods rather than anything particularly new.

It can be said therefore that we already have the technology we need to extract *conventional* oil efficiently. Indeed, we have had this capability for many years, meaning that the record of the past is already the beneficiary of modern technology with little scope for further improvement. The constraints are not technical but imposed by nature.

It is also claimed that technology will recover a higher percentage of the oil from the reservoir. This also does not bear close analysis. A 30 per cent recovery factor was routinely applied in earlier years. It is said that the improvement to the current 40 per cent average reflects technological advances. In fact, the 30 per cent was little more than a rule-of-thumb because few fields had then been abandoned and no one knew what they would deliver. Besides, the amount of oil-in-place on which the recovery factor is based could not be accurately determined with the relatively primitive technology then available. The estimate of recovery in a mature field is based primarily on the well performance, and no one particularly cares what percentage this might be of a notional value of oil-in-place.

In short, we can abandon hopes that technology will release vast amounts of new *conventional* oil. Most of the improvements that have been attributed to technology are in fact the consequence of commercial or regulatory under-reporting.

Production scenarios

Oil is most unevenly distributed as illustrated in Figure 3.7. North America has used most of its endowment. About half of what remains to produce lies in just five Middle East countries (Abu Dhabi, Iran, Iraq, Kuwait and Saudi Arabia). This situation is a consequence of the expropriations of the 1970s when the major companies lost control of their principal sources of supply as they found themselves the unwelcome tenants of increasingly unfriendly landlords. Had they remained in control, they would have produced the relatively cheap and easy oil of the Middle East before turning to the progressively more costly and difficult provinces offshore and in Alaska. Market signals of impending shortage would, therefore, have been delivered gradually as the industry was forced to move to the more costly and difficult sources. This distortion is contrary to the normal economic pattern. It has imposed a certain *swing* role on the five Middle East countries, which can use this cheap and easy oil – at least for a limited period of time around peak – to make up the difference between world demand and what the rest of the world can produce within its resource constraints. We have to take this factor into account in modelling future production. The *swing share* is the proportion of total output produced by the above five Middle East countries.

A simple model of production that respects the endowment in nature and the fact that peak has to come close to the midpoint of depletion can be developed from the information provided in Table 3.1. Production in a country that has passed its depletion midpoint can be assumed to decline at its current depletion rate, namely annual production as a percentage of the *yet-to-produce*. Production in a country not yet at midpoint can be expected to rise to midpoint and then deplete at the midpoint depletion rate. Since most such countries are close to midpoint, the assumptions about the rate of increase are not critical, and a rounded 5 per cent annual increase will suffice in the absence of any other insight.

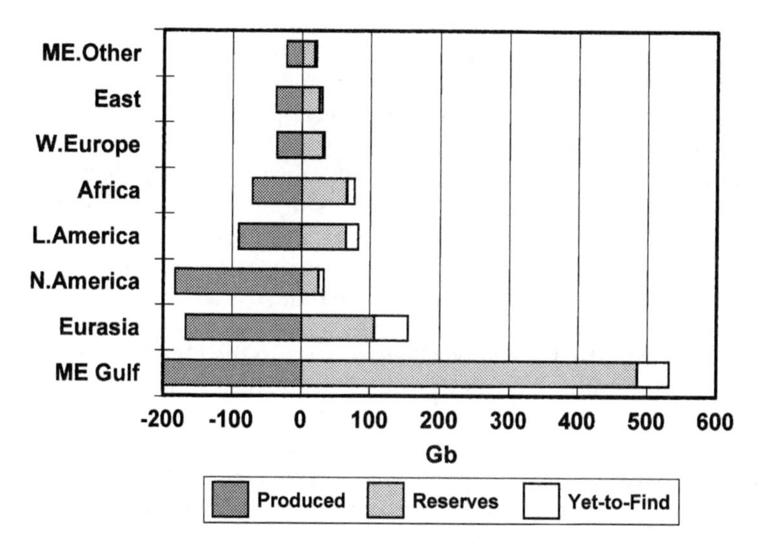

Figure 3.7 Distribution of oil. North America has used most of its oil and the Middle East has most left

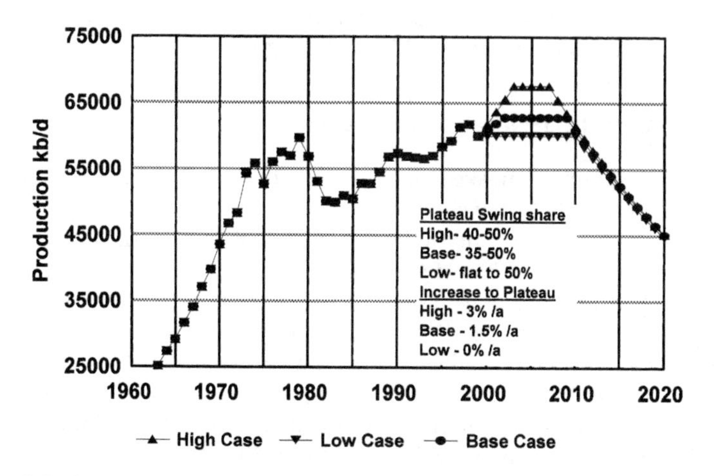

Figure 3.8 Scenarios of world production. Recent events suggest that the 'low case' is more likely than the 'base case' as a result of world recession

Figure 3.8 shows three scenarios that have been proposed to span the range of realistic possibility.

1. *High case* – this assumes that production rises at 3 per cent a year in a world enjoying great economic prosperity until the five Middle East Swing Countries produce 40 per cent of the world's oil. That control is taken to prompt a leap in price which curbs further increases in demand giving a plateau of production. It lasts until swing share reaches 50 per cent, by which time these countries are approaching their depletion mid-point, and production is assumed to decline at the then depletion rate of about three per cent a year.
2. *Base case* – this is the same as the *high case,* save that production rises at 1.5 per cent a year in a less prosperous world to the plateau of production commencing when *swing share* reaches 35 per cent.
3. *Low case* – this is also the same but assumes no increase in demand, assuming economic recession or stagnation.

The *swing share* itself has to be allocated back to the individual Middle East countries, including the special case of Iraq, which has as consequence of the Gulf War become a swing producer of last recourse. Its exports were at first subject to an embargo, which was later relaxed when oil prices rose (on a humanitarian pretext). Iraq's oil will soon be desperately needed, and with the boot on the other foot, Iraq is unlikely to be sympathetic to Western demands, carrying risks that new pretexts may be found for Western military intervention.

These are long term scenarios, from which temporary departures for all sorts of transient reasons are to be expected. They do, however, to some degree self-adjust, because if production is higher than modelled for a particular year, the depletion rate will also be higher.

The anomaly of 1998 and the reaction

Non-swing production peaked and began to decline in 1998. It would therefore have been logical to expect prices to have risen as increasing control of the market passed to the Middle East swing countries, but the opposite happened because of a number of extraneous factors including warm weather, an Asian recession, the devaluation of the Russian rouble that encouraged exports, and the under-reporting of the supply by the International Energy Agency which prompted OPEC to increase production. We may also be forgiven for wondering if the market had been manipulated by major companies and their bankers talking down the price of oil to facilitate the major acquisitions they were making. The market is driven by the marginal barrel and is very sensitive to sentiment and facile comment.

Oil prices (Brent Crude) fell to a low of about US$10 per barrel in December 1998, only to rebound three-fold within 18 months. As the market surged through US$30 in February 2000, the United States government began to apply pressure on OPEC to increase production, and the market started marking down the price in anticipation of an OPEC meeting. OPEC meanwhile dispatched a number of cargoes, some drawn from floating storage in the Gulf rather than production. Their arrival was timed to coincide with their March meeting, at which they offered to try to hold the price in the US$22–28 range. The US Geological Survey did its bit by making a press release of an unissued report claiming that the world was awash with oil. By April, prices had dropped to US$22, but the mini-glut was soon gone, and prices rebounded – first to test the OPEC ceiling of US$28 and then the emotional barrier of US$30.

Much discussion addressed the issue of spare capacity. Some analysts, ignoring depletion, wrongly assuming that spare capacity was simply the difference between a country's current production and its historical maximum. Spare capacity can mean many things. Opening a free-flowing well that had been deliberately closed releases the corresponding capacity immediately, but other measures such as infill drilling, developing so called mothballed fields (if any), and exploration, all take investment, work and, above all, time to deliver. Gradually it became apparent that there was in fact virtually no spare operational capacity as countries desperately tried to offset the natural decline of their old fields holding most of their oil. Production in the United Kingdom for example fell 7 per cent in 2000 (and 16 per cent October to October) despite every heroic effort. With the US election approaching, President Clinton decided to release oil from the Strategic Reserve to calm the market as fears of a fuel crisis spread through the United States and as protestors against high fuel costs marched in Europe, bringing Britain to a standstill for a few days.

These actions combined with a certain backdown by Iraq to suspend exports, and the liquidation of speculative positions for year-end financial reporting purposes, led to another anomalous fall in December 2000. It would have been logical to expect upward price movements to continue as depletion bit ever harder, but in fact the price rises triggered an economic recession, cutting demand and thereby reducing pressure on price. It is impossible to predict the short term responses to these events, but it seems clear that the market will sooner or later perceive that there is in fact no ceiling to oil price save for curbing demand. Figure 3.9 shows the relationship between swing share and price.

The inescapable conclusion that only falls in demand can curb escalation in price is a devastating one, speaking of world recession and a plunging stock-market, of which at the time of writing, there appear to be serious fears, especially since any economic upturn would repeat the vicious circle

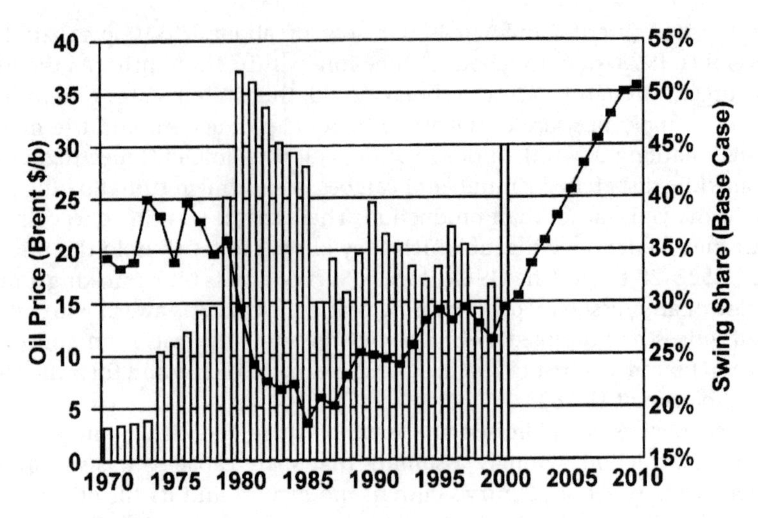

Figure 3.9 Swing (Middle East) share and price

of rising demand and price. It is not surprising that governments are so reluctant to face up to the situation. They are reluctant to give warnings for fear that they may prove to be self-fulfilling, and they realise that if they admit to having foreseen the event, they may be held responsible for not having made appropriate preparations. In political terms, it is easier to let the crisis be seen as an act of God and hope to win votes by reacting to it.

Conventional oil is not the only hydrocarbon fuel and feedstock, important as it is. Gas and non-conventional hydrocarbons will become increasingly important after peak production, having their own depletion characteristics.

Gas and non-conventional hydrocarbons

Gas

Gas was much more widely generated in Nature than was oil but its smaller molecular size, giving it greater mobility, has several important consequences. First, it leaks more easily from traps; secondly, a higher percentage, normally around 75 per cent, is recoverable; and thirdly, its depletion profile is normally capped by market and facilities, giving constant supply over a long period coming to an abrupt end. In effect, the depletion profile provides a huge hidden balloon of spare capacity of very cheap gas, which is progressively drawn down in an unregulated market. Prices continue to fall in apparent vindication of free market pressures until the end of the plateau on the exhaustion of the spare capacity when production plummets and prices soar without warning (Udall and Andrews, 2001).

Gas is much more costly than oil to transport, meaning that whereas there is a global market for oil, gas belongs to three separate major markets in North America, Western Europe and Southeast Asia. There are also large deposits of stranded gas, so far lacking an outlet, in the Arctic regions of Canada, Alaska and Siberia, as well as in Nigeria and to a degree in the Middle East.

Digging further into detail we find new problems of definition and measurement. What is an oilfield and what is a gasfield? Oilfields commonly contain expanding gas caps, so that at the end of their lives, they may hold more gas than oil. Gas is also commonly reinjected, partly to avoid flaring in remote areas and partly to pressurize the oil reservoir. Gas was previously commonly flared as an unwanted by-product of oil, and often contains large amounts of inert gas, especially carbon dioxide. The data do not always make clear what is reported. Furthermore, there is the issue of equivalence with oil, being variously based on value, volume or calorific content. For general purposes, remembering the weakness of the input data, it is convenient to equate one billion barrels of oil with ten trillion cubic feet of gas, the customary units of measurement.

Gas is much less depleted than is oil. The essential statistics given below, as of end 1999, are less sure than is the case for oil.

Trillion cubic feet	
Produced in 1999	90
Produced-to-date	2,300
Reserves	5,700
Discovered	8,000
Yet-to-find	2,000
Yet-to-produce	10,000

It is also much more difficult to model depletion, because the profile is dictated more by market forces than the immutable physics of the reservoir, which should not, however, be ignored. Whereas oil depletion follows a general bell-curve, world gas production is likely to rise in a series of steps reflecting new pipe line linkages, especially to the Middle East. The midpoint of depletion, which probably still approximates with peak, will likely come around 2020.

It is worth noting however that North America faces a dire gas supply situation. In fact California is fast running out of spare electricity generating capacity leading to rolling black-outs, and the same problem is likely to spread to other states. The United States is trying to make good the shortfall by importing Canadian gas, only to find that Canadian reserves too are being rapidly depleted. New pipelines to non-conventional Arctic gas will have to be built as a matter of urgency and liquified gas imports stepped up, but they are unlikely to meet demand that has been stimulated by the lower prices of the past and to some extent by environmental concerns. It comes as a shock to realise that the famous US economic miracle may be over.

Europe faces the same general future with regard to its indigenous supply, but has the advantage of being able to tap the substantial reserves of the former Soviet Union, the Middle East and North Africa, despite the dangerous new geopolitical dependency that implies.

Natural gas liquids (NGL) and oil from coal and 'shale'

Gas contains dissolved liquid hydrocarbons, especially where it is derived from overheated oil in the basin depths. Some of this liquid condenses naturally when brought to the surface, being termed *condensate*, and more may be extracted by processing.

Yet again, we face severe difficulties of measurement and definition. The condensate may or may not be metered together with the oil, whereas the NGL from processing is normally reported separately. In principle, NGLs should be treated as part of the gas domain, but in practice they are commonly included in the oil statistics of production and reserves, being sometimes collectively termed *liquids*. This is yet another source of confusion, likely to become worse as more gas is produced. As a rule of thumb 1 trillion cubic feet of gas yields 13 million barrels of condensate and, with processing, up to 40 million barrels.

New technology is being developed to convert gas itself into liquids, using a variant of the Fischer-Tropff method that converts coal into liquid hydrocarbons, as was used by Germany during the Second World War and South Africa during the fuel embargo. It is known under the acronym GTL. So far, only two sub-commercial pilot plants are in operation, but there is every possibility that production will grow radically as oil prices soar.

Immature source-rocks, confusingly termed 'oil shales', can also be retorted at high temperatures to give up oil in a process similar to that of nature in the case of conventional oil. The retorting consumes almost as much energy as it delivers, and there are environmental constraints due to the consumption of water and the fact that the waste product is a toxic powder. Oil shales have been exploited on a small scale since the Middle Ages, but the efforts to undertake large scale extraction, particularly in the United States after the oil shocks of the 1970s, have all come to nought. The resource is large, and extraction on a local basis will probably increase as conventional oil becomes scarce and expensive, but in no sense can it be considered a substitute.

Tar and extra-heavy oil

One of the factors that makes the Middle East such a prolific oil basin is the widespread occurrence of salt that has sealed the oil traps. Other areas of prolific generation were not so endowed, and the oil migrated to shallow depths on the margins of the basins where it was weathered and attacked by bacteria, leaving behind huge deposits of tar and extra heavy oil.[3] The two largest deposits are in Eastern Venezuela and Western Canada, but there are others including some in Russia.

In Canada, the deposit is mined in huge open-cast pits after the removal of as much as 75 metres of overburden. The ore – for that is what it effectively is – is then centrifuged and processed using cheap stranded gas as a fuel to yield a high quality, sulphur free synthetic oil. In Venezuela, the deposit lies at greater depth and has been extracted by drilling patterns of five closely spaced wells, using the peripheral wells for steam injection to mobilise the oil and drive it to the central well. More modern methods are now being applied in both areas by drilling twinned long-reach horizontal wells for, respectively, steam injection and recovery.

It is important to remember that these are not homogeneous deposits. So far, only the most favourable locations have been exploited, meaning that further developments become progressively more difficult and costly, also in environmental terms. Mention may also be made of *orimulsion*, an emulsified heavy oil manufactured in Venezuela, that has been used directly as a bunker fuel, although its high sulphur content has limited its market for environmental reasons.

Approximately 1.7 mb/d are currently produced from these sources, and production will no doubt be stepped up in the future, rising perhaps to as much as 4 mb/d by 2020. The resource is huge, so that rate of production can be sustained for a long time.

Heavy oil

The extra-heavy oils referred to above grade into heavy oils, which in turn grade into conventional oils. But, as by now we may expect, there is no clear definition of the boundaries. In Venezuela, the cutoff is taken at a density of 22° API, whereas in Canada it is even higher at 25° API. It is an arbitrary cutoff in any case. Here it is placed at 17.5° API so that most (but not all) producing heavy oil fields are treated as conventional.

Polar oil and gas

The polar areas are here treated as non-conventional not only because of the extremely hostile environment, but because the geology is rather different and our knowledge is less. Generally speaking, Antarctica does not appear to be very prospective, being in any case closed to exploration for environmental reasons, by international agreement. The Arctic, which contains some very large sedimentary basins, is more promising, but the evidence to date suggests that it is generally gas prone, probably because large vertical movements of the crust under fluctuating ice caps in the geological past have adversely affected the source-rocks. Alaska, which does contain normal oil, generated in the Triassic under unique conditions, is something of an anomaly.

Deepwater oil and gas

No one would look for oil in the ocean depths were there anywhere else easier left. Necessity is nevertheless the mother of invention, and methods to undertake deepwater operations have been developed. While now feasible, they remain at the limit of what can be done. The geology of the main prospective deepwater domains differs markedly from the normal continental shelves. The most prolific areas are found in the Gulf of Mexico and on the margins of the South Atlantic, where early rifts in divergent plate tectonic settings contain effective oil source rocks, generating oil that has locally migrated upwards to fill slumped reservoirs, which have in turn been winnowed by long-shore currents. In other places, deltas have extended into deep water but, in the absence of underlying generating rifts, are likely to be gas prone with only minor oil accumulations. On present evidence, it looks as if the deepwater domain might contain about 50–85 billion barrels, and an as yet unquantified amount of gas.

Conclusions

Having now outlined the resource base and the general depletion profile imposed by nature, we may turn attention to some of the possible consequences. It is a large subject, and we cannot here hope to do more than touch on some of the issues in summary fashion.

We may distinguish scenarios in two broad groups of respectively growing and falling oil demand, reflecting economic growth or decline. If demand rises, prices must soar because of the resource limitations and growing dependence on the Middle East, an event that would concentrate the minds of governments wonderfully. Initially OPEC will no doubt be accused of holding the world to ransom and all sorts of pressures will be applied. But gradually it will become evident that OPEC cannot rapidly increase production even if it wanted to as it is already running flat out, trying to offset the natural decline of its giant old fields holding most of their oil. The West has every incentive to bury the hatchet with Iraq and work hard to extend a new hand of friendship, but the renewed bombing of Iraq in early 2001 may be an ugly portent of what is to come from misguided policies.

The United States will have to come to terms with its own critical gas shortage for which there is no solution in terms of added domestic supply. The people at large will become very much aware of the energy situation, and gradually governments and oil companies will be forced out of denial to address and explain what the real position is. Already Shell advertisements emphasise that the world needs the renewable energy Shell is investing in, implying that oil is insufficient. The chief executive officer of BP (formerly British Petroleum) confessed at the Davos meeting that maximum output at 90 mb/d would be reached by 2010, having previously changed the company's logo to a sunflower, saying that BP stood for 'beyond petroleum'. Exxon–Mobil stresses the extreme difficulty of offsetting depletion, which it says is running at two to three per cent per year in OPEC and seven to eight per cent per year in non-OPEC countries. Governments and corporations may be surprised at how positively people react once they are properly informed.

If on the other hand demand stabilises or falls from growing economic recession, there is less pressure on prices and the peak of production is delayed. Furthermore, less demand will reduce the incentive to step up exploration, bring in non-conventionals, as well as renewable energy and energy saving. But the elasticity in demand is limited. Even if we park the Mercedes, we will still need fuel for the tractor. Accordingly, the imbalance between demand and supply will sooner or later reappear, pushing prices upward.

This epoch of transition will probably last about ten years and will undoubtedly be one of great tension and readjustment, but gradually the world will come to accept that radical changes are being forced upon it. Global trade will dwindle and self-sufficiency will become a priority. The United States and, to a lesser degree, Europe will move into deep recession as their economies built on consumerism grind to a stop. People may again darn socks, instead of importing new ones from Taiwan, and they might even enjoy doing so. The scale of the transition is daunting.

Nuclear energy will probably be rediscovered with improving safety, and the full spectrum of renewable energy will receive emphasis. Energy saving everywhere, for which there is colossal scope, will be a priority. Airlines will all but crash. Public transport will be the norm. Town planning will have energy considerations as a priority, moving towards the concept of miniaturisation and self-sufficient units. Building codes will change. The list is endless, and it is hard to speculate on what exactly this turning point will bring, but that it is a turning point can hardly be doubted.

Notes

1. Reserve reporting is a subject to itself. Modern methods apply statistical probability theory identifying 95 per cent probability, 5 per cent probability, Mode, Mean and Median values in a log-normal distribution, sometimes further refined by Monte Carlo simulation. In practice, this is more impressive in appearance than substance because the input is so unreliable. In general, it seems better to aim at a single best estimate near the centre of the range of uncertainty, such that revisions are statistically neutral.

2. Here we define *Ultimate as Cumulative Production* by 2075 to avoid having to worry about irrelevant tail-end production having no impact on peak, our primary concern, but it makes no practical difference.
3. Tar, which is bitumen, is defined on viscosity, whereas Extra-Heavy oil is defined on its density (< 10° API).

4

The Effect of e-Commerce on Transportation Demand

Hartmut Stiller

Introduction

The most remarkable change in the world economy of the past decade has been the explosive increase of information technology (IT), first of all the Internet. Notwithstanding the current disillusionment with dot com companies, the Internet is seen as one of the key technologies of the 21st century. In consequence there are many issues which might affect transport demand and sustainability, ranging from the impact of changes in work and personal relations up to new ways in politics and the handling of conflicts. This chapter will pay attention primarily to the impact of e-commerce on transport demand and only briefly touch on some of the other points.

The reason why transport is under discussion is not transport itself but rather its impact on sustainability, especially the environmental impact of road and air transport. Thus, in this chapter a perspective will be presented of resource productivity as a measure of efficiency in the use of natural resources and related CO_2 emissions.

The development of transport demand in the 1990s has not been very promising with regard to its impact on sustainable development. There has been a strong increase in traffic volume on a national and also on a global scale and a (further) shift to more unsustainble modes of transport. Improvements in fuel efficiency and emission parameters has been more than absorbed by growth in volume in most areas. In this respect, hopes are being placed in the Internet to bring these environmentally unsustainable trends to a halt and lead to a decoupling of transport demand and growth. Thus it has to be asked whether the so-called 'new economy' will be more ecologically sustainable than the 'old' or whether there will be a repetition of the pattern of the past when new technologies have been introduced, namely that efficiency gains are absorbed by new additional services.

As communication is revolutionised, the Internet offers consumers, companies, and governments (stakeholders) new ways to express and channel their preferences and products which should improve market efficiency.

There is the hope that e-commerce will improve the efficiency of markets and virtualise many products and services and will allow business models based on 'access' instead of 'property' (Rifkin, 2000), thus reducing transport demand and improving resource efficiency.[1] So optimists assume that IT might even accelerate the trend continuously to dematerialise our GDP (Adriaanse *et al.*, 1995, 1997; Hawken *et al.*, 1999). Today the world produces nearly twice as many services as twenty years ago while using roughly the same amount of resources. On the other hand, up to now there is no evidence that rebound effects will not water down these specific efficiency gains. Rather the energy crisis in California in 2001, for example, suggests that electricity consumption will continue rising to keep pace with or even to exceed the growth of GDP.

In the past decade transport volume has grown by 50 per cent on a global scale. However the Internet appeared only in the second half of the nineties. The question is how much did it contribute to the growth in transport? What can we expect for the future? Will it even *accelerate* transport growth or will it finally lead to a reduction in transport demand?

The main thesis of this chapter is that the direct impact of e-commerce on transport demand will be limited, accounting for less than 10 per cent change in transport volume. In this respect e-commerce is defined as ordering or selling of products or services using the Internet. E-commerce might thus imply that the product itself is provided or just traded electronically. Thus, an assessment has especially to focus on the following questions:

1. How much does the Internet increase the efficiency of the markets for commodities and transport thus, *ceteris paribus*, reducing transport demand?
2. How much does the Internet reduce transport volume by digitalisation?
3. How far will e-commerce lead to growth in the average distance travelled?
4. Are there structural effects in the economy that could lead to a reduction in transport demand through e-commerce and the Internet?
5. What about growth-induced increase in transport demand?
6. What are the direct and indirect impacts on passenger transport?

The economic system is too complex to permit a definite answer to these questions. However, this chapter will try to identify the crucial factors. Transport demand in different industrial sectors will briefly be analysed and the results combined to arrive at a comprehensive scenario about the impact of e-commerce on transport demand. Similarly, changes in passenger transport demand are also discussed. To underline the theoretical findings, the impact of 'business to consumer' electronic commerce (B2C) on the transport demand for the delivery of books will be briefly analysed as an example. This example also shows the need for a common denominator for both freight and passenger services as e-commerce might alter demand for both. Here gross transport volume is taken to allocate the impact of e-commerce using transport demand, resource productivity and accumulated energy as key sustainability indicators.

B2B (business to business)

E-commerce – the current situation

Currently the bulk of e-commerce takes place between companies. B2B is introduced to make procurement more efficient, to reduce transaction costs, and to enlarge the geographical range of possible suppliers. There is evidence that e-business leads to a reduction of inventories and stocks which improves the efficiency of production.[2] This development seems to be environmentally desirable, as huge inventories tend to end up in larger waste volumes of goods which are no longer marketable. So B2B can reduce the overall waste volume. Sometimes reduced inventories imply a higher transport frequency, often done by less environment-friendly transport modes such as air transport. Primarily, the Internet allows an intensified exchange of information along the supply chain.

Exchanges

The most important innovation of e-commerce is the emergence of electronic market places. The economic idea of electronic exchanges is obvious: greater market transparency and larger geographical scope, as well as much smaller transaction costs leading to cost reductions in procurement, and efficiency gains along the supply chain. The success of electronic exchanges depends to a large extent on how readily a product can be standardised. If there is only a limited number of classes or types of product, it is easy for buyers to change suppliers. As more specific product features are essential – durability of a part, surface quality and so on, markets become thinner and the advantages of the Internet for e-business tend to disappear.

The volume of commodities suitable for B2B-commerce – apart from the primary materials already discussed above – is, in principle, very large. However, standardisation is not the only point. If there is a reduction in quality of products bought on an electronic market place, endangering the quality of the *brand*, the savings in general will not compensate for the risks. In those cases B2B-exchanges might work as information exchanges or as a price indicator but, in the end, buyers will opt for durable supply chain relations. Therefore, it is not surprising that today B2B-exchanges strive to add other services for their clients as they check the integrity of supplies or even certify them. In this case, however, the exchange agent might end up as an information trader itself, collecting data on supply and demand.

At present more than a thousand product and industry specific exchanges exist. Even if a market consolidation actually takes place, it is assumed that a growing percentage of all products will be traded on such exchanges.

Impact on transport demand

The impact of B2B on transport demand is not so easy to calculate. However, several factors call for an increase in transportation demand which is not the result of e-commerce alone. For years companies have tried to reduce their stocks and inventories. Nevertheless, B2B might be helpful in this respect as it enhances market transparency and allows companies to reduce their security stocks (reserves necessary to permit production to continue in the event of an unexpected interruption to the supply chain). However, as long as the production structure remains unchanged, the potential for changes in transport are limited.

In the long run, the mode of procurement itself may change. Replenishment is already used by big production companies in some areas so that regional wholesalers can concentrate on selling products to retailers and customers. Nonetheless, the direct impact in the trade of bulk commodities due to the Internet is rather limited. All things being equal, the emergence of the Internet should not change the trade structures and prices for most primary materials. Spot markets such as the Rotterdam market for oil already existed before the Internet emerged. Thus, this part of world trade should not be very much changed by the opportunities of trading online.

However, the Internet might *indirectly* affect those markets in a serious way resulting in significant changes in demand in general and so also in transport demand. For example, if the Internet and e-commerce were to result in a reduction in paper demand the consequences for the timber, pulp and paper markets will be obvious. Changing transport relations resulting from changes in the market structure and volume are beyond the scope of this chapter. However, it is rather unlikely that a change in volume alone will deeply transform markets. Rather it might accelerate changes due to distortions that already exist. Thus, transport demand caused by those commodities as a first order effect will be unchanged. In addition, at least from the environmental point of view, the most decisive impact of e-commerce is on road and air transport. So in this respect, the development of the markets for primary materials is not of great relevance.

The impact on sustainability as well as the environmental consequences (impact) of approaching perfect markets are unclear. In general, such markets tend to be more efficient. Thus, they should allocate resources as well as natural resources in a better way. In theory, if there are no externalities, B2B might enhance weak sustainability as it improves resource efficiency. Yet, in this case, from the pure environmental point of view, such a more efficient allocation of resources does not necessarily lead to ecologically favourable results. That is the case if the more efficient use of some production factors also implies an enlarged use of the environment as a production factor.

However, the basic assumption of this conclusion rarely holds as full costs are not born by transport prices. As long as environmental resources are only partly allocated efficiently and treated as common goods, and as long as specific environmentally relevant activities such as certain modes of transport are subsidised, the overall impact of B2B on the environment may easily be negative. Moreover, things might even get worse. In an imperfect market with imperfections in several production factors, a reduction of the distortion in one production factor does not necessarily improve overall wealth. If improved transparency and a larger scope of e-exchanges are accompanied by a strong increase in subsidies, the result for transport could be an increase of the official GDP and at the same time a reduction of overall wealth and weak sustainability.

The impact of B2B exchanges on transport distance

The extension of geographical scope is not a big issue in B2B. Procurement in many companies has always been dominated by concerns about costs while geographical restrictions have been of minor importance. Therefore, the effect on additional transport and longer transport distances might not be severe. In many cases, however, because purchasers in industry just don't know very much at present about many international sources of supply and their prices, electronic platforms could push further internationalisation.

Enhancing market transparency will favour supplies located at a greater distance. Often it's no coincidence if companies are supplied over a long distance. In those cases typically either an advanced product quality or price advantages exist (or at least existed) or trade, in effect, is intra-company trade. At the moment, companies generally know much more about local supplies than those far away. Here lies a significant potential for increasing transport demand. As long as transport prices are subsidised this can easily go beyond economic efficiency.

Smaller order size

More frightening are the environmental consequences of the dramatic reduction in transaction costs in some areas. Companies are experimenting with purchasing models that will allow their employees to order their office materials individually via the Internet from certain pre-selected suppliers. The direct delivery of these items to the workplace implies smaller orders. So we will face more deliveries and more transport on a small scale and more packaging. High economic costs of warehouse, handling and administration will be replaced by additional transport costs in parcel services which do not include the full environmental externalities.

The impact on modal split

Besides influencing the total transportation demand there may also be a shift in transport modes towards road and air transport which could have a negative impact on sustainability. Competition in B2B exchange will emphasise not only the price but also short delivery time. To compete with local competitors, global (or inter-regional) suppliers will strive to reduce their delivery time to every destination. This will lead to a general acceleration of the speed of moving goods to market. Pressure on delivery times will tend to favour air and road transport which, in turn, might lead to a further increase in road and air transport in B2B exchanges.

The impact on transport markets

Finally, e-commerce has also changed transport markets. Electronic exchanges for excess transport capacity are meant to allow a more efficient use of vehicles, and therefore, to reduce the specific environmental impact of a single unit of transport service. Currently about 24 per cent of all capacity travels empty on German roads but the overall capacity use in transport performance (T-Leistung) is only about 46 per cent (German Office for Transport). If transport exchanges only helped to increase this figure by a few percentage points, e-commerce would have a positive impact on sustainability. However, up to now, German transport statistics show two contradictory trends in this respect. The number of empty trips has been falling slightly – from 44.1 per cent to 41 per cent since 1995. However, at the same time the average use of full load capacity has also been falling – from 64.4 per cent to 59.8 per cent, leaving the overall efficiency of total use of freight capacity nearly unchanged (around 46 per cent).

Partly this fall may result from the decline in the German construction industry (which has an average freight capacity use of 72 per cent), but this does not explain the development completely. The rest is up for pure speculation. One explanation may be that there is a trend to lighter products, a second that there is an increasing need to deliver in smaller quantities, and a third that a reduction in stocks is leading to similar effects.

Transport statistics may be telling us that transport exchanges indeed enhance transport efficiency by reducing the amount of empty trips. On the other hand in 1999 the last restrictions on the European single market in transport were lifted, allowing foreign trucks and transport companies to offer their services within the entire European Union, contributing to increased movement. Moreover, efficiency gains may be absorbed at least in part by rebound effects. In competitive transport markets it is unlikely that efficiency gains will enhance profit margins. More likely falling transport prices will induce additional demand for transport services.

All in all, the figures tend to indicate that up to now the structural changes resulting from information technology do *not* enhance the efficiency of the transport market. Therefore, to assess the impact of the Internet on transportation demand, the effect on volumes and on distances in freight transport has to be analysed.

B2C (business to consumer)

Characteristics

Most of the points for B2B apply also to B2C. However, there are some differences in a few crucial aspects which are also of environmental relevance. One distinguishing factor is the size of business transactions. Consumers rarely spend more than 1000 Euro at a time (except for cars), but instead buy many items for less than 50 or even 10 Euro. So transaction and transport costs are of greater importance than in B2B.

Compared with business exchanges the Internet dramatically enlarges the scope of products offered to consumers. Opportunities for consumers are traditionally limited to products they can buy locally or via the catalogues of a few mail order houses. All other products have to be bought indirectly via a retailer. Now the Internet offers consumers the opportunity to shop globally. The direct environmental consequence is a sharp increase in air freight and volume for parcel services. Although the volume of such purchases is small at present there is much potential to cannibalise the exuberant costs of retailing.

The Internet offers a huge advantage for consumers in transaction costs. The Internet has the potential to provide much better opportunities for consumers to select the product desired. But when it comes to transport, the most costly part of the supply chain is actually assigned to consumers. People invest a lot of time in packing and taking home their shopped items. Especially in food retailing, this shifting of activities on to consumers is a crucial factor in the intense price competition among supermarkets. So, to challenge the traditional food retailing sector requires either a change of people's attitudes towards price and comfort or the formation of more cost-efficient delivery structures. Although these aspects are most relevant for food retailing, they apply also to other consumer products.

Unfortunately, simple solutions to bridge 'the last mile' (final distribution to the consumer) are not in sight. Environmentally speaking, transport for shopping is a significant part of all urban transport. If B2C were to replace many shopping trips by an organised delivery system, transport volume and impact could be significantly reduced. However, it is uncertain what people would do with the spare time gained. As long as distances for shopping are short, at least for daily products, then the danger exists that people would use the time gained for other trips further from home which, in the end, could even result in an overall increase in passenger transport.

The potential of electronic trade

In principle, many goods can be ordered electronically. The existence of mail order trade shows that even in the past, long before the Internet emerged, many goods were sold directly to consumers. A look in a catalogue shows that clothes, china, lamps, furniture and so on can be sold successfully without a retail shop. And goods comprising information – music, video, newspapers, magazines, banking services and so on – are among the most successful items traded electronically.

The economic advantage of e-commerce results primarily in lower costs for retail space and lower personal costs. In many shops the labour costs of sales staff make up a large part of the costs without adding much value for the consumer. Up to now, at least in Europe, there have been few opportunites for the consumer to avoid these costs. The Internet will challenge employment in retail sales. Although they are not well paid, the size of the workforce in shopping malls and shops is enormous. Either sales staff will have to offer consumers additional services they are willing to pay for, or those jobs will sooner or later disappear.

Whether food can be sold successfully over the Internet remains to be seen. A first attempt in California failed, and the business went bust when the hype was over. Consumers will always insist on looking over items such as fresh fish or fruit before buying. On the other hand, why not shop at Aldi or Walmart (supermarkets) electronically? The development in this field also has a huge potential impact on transport demand. Why the prospects of e-commerce in this area are not so promising will become clear in the following sections.

Obstacles

Delivery

For e-commerce to replace traditional commerce there must be either clear economic advantage for consumers or a major additional service advantage. Whether home delivery represents such an advantage is yet to be proved since it is the most expensive way of transferring the purchased good to the consumer. In the past twenty or thirty years there has been a strong trend to reduce prices by shifting certain purchase costs to the consumer: take off-the-shelf in supermarkets, 'bring back' systems of shopping trolleys or weighing of goods. Even in e-commerce such shifts can be observed: lower charges for direct banking are attributable in no small part to client self-service.

It seems highly unlikely that many consumers – especially those with a small budget – would be willing to pay for extra services which they can easily undertake themselves. Therefore, home delivery is attractive mainly for business people and those who are less mobile like pensioners or handicapped people. However, in both cases there may be obstacles. Pensioners are less accustomed to surfing the Internet, and business people often spend only a few hours at home and need somebody to receive delivered items. So the market for home delivery may be limited. Rather business models that allow the consumer to harvest part of the cost saving can be expected to be most successful.[3] One idea is shopping points where people can take away their electronically ordered goods. This would combine the cost saving potential of shops but shift the bulk of transport costs for the last mile to the consumer. However, if those places are to be close enough to the consumers home, e-commerce must take a high market share. If there are only a few points within a city, the time savings will melt away.

The impact of home delivery on transport demand is twofold: On the one hand, transport by vans will grow. On the other hand, home delivery will reduce private trips for shopping. Currently, in Germany, shopping is responsible for about 12 per cent of all passenger transport. The transport efficiency of combined home delivery is much better than single private shopping trips. In extreme cases a car weighing more than a ton is moved to buy a single item weighing a few hundred grams. A van – even if it is 80 per cent empty – will carry more than a hundred kilograms of goods. The much larger average freight volume will more than compensate, by a large margin, for the longer delivery trips and the slightly higher specific fuel consumption of the vehicles. Most models of e-commerce assume home delivery of shopped items. The last mile is the crucial factor for e-commerce.

Brands and trust

The consumer's trust in the quality of a product or a service is a crucial factor for successful business. Here e-business faces a serious disadvantage because the consumer misses the physical shopping experience – an important event where trust is created. Over time, however, experience and satisfaction with previous orders can reduce such a factor. Brands – either product brands such as Coca Cola or brands of retailers such as H & M – are an important guide for consumers. So it is rather unlikely that many people will buy, for example, cheaper shirts of the same quality from East Asia via the Internet than H & M carries in its stores. In consequence, the Internet will probably not lead to the disappearance of traders such as Metro or Marks & Spencer. Rather those companies are best placed to take the lead in e-commerce because they have knowledge of supplier markets around the globe, the logistics and power to set attractive prices and the brand consumers know and trust.

In theory, companies with brand products could try to sell directly to consumers, bypassing retailers and adding retail margins to their own profits. However, there is a serious economic obstacle in transforming an existing brand into a pure e-brand. If the e-brand is to be successful, it must show a price advantage. However, such a price advantage could easily reduce profits. Without a significant price advantage the shift towards e-commerce will be rather slow and closing down ordinary shops is no advantage if it reduces sales volumes significantly. Hence it is unlikely that brands will speed up the shift towards e-commerce for economic reasons even if it were to be more cost-efficient than traditional retailing.

Because of the need to create a brand, Internet companies have invested heavily in developing an image consumers know and trust. However, only three of the thousands of Internet companies have been successful in creating a global brand: Yahoo, Amazon and ebay (see: 'Internet Pioneers'

The Economist, 3 February, 2001: 79–81). And only Amazon has been relatively successful in competing directly with traditional retail companies. In most cases, *traditional* brands are used for e-business. If the most successful companies in e-commerce are traditional 'brick and mortar' companies, transport demand along the supply chain to the wholesaler will not change very much. E-commerce will be just an additional channel to reach consumers. Products will either be delivered from the same distribution centres for all retail stores or by using existing logistics to reach those stores and just adding final distribution within a city. The impact on transport demand of the upper side of the supply chain thus depends on the efficiency advantages of logistics companies in respect to company/wholesaler logistic networks. In the past few years the liberalisation of transport markets has allowed the replacement of inefficient intra-company logistics. If the remaining company logistics is as efficient as that on offer by logistics companies, e-commerce will not have a large effect on transport demand in this part of the supply chain.

Culture

A further argument against e-commerce taking a huge market share in the near future is that such a change would require dramatic cultural changes in societies. Serious efforts have been to make older people enthusiastic about the Internet. However, resistance to the use of credit cards, security problems, and shopping habits as well as customer loyality are reasons why a rapid shift towards e-commerce is unlikely. Why invest in high information costs and take pioneer risks to find a reliable online-shop if you know an established store around the corner? At least for those products where there is no significant price advantage in buying them over the Internet – and this seems to be the vast majority of cases – consumer behaviour is economically rational in maintaining established supplier relations.

Social learning is not easy. At present mail order business has a market share in retail in Germany of about 10 per cent. Why has it been rather constant over time? Well, family experience could be one explanation. In consequence, why should the Internet be much more successful? Of course, the young will try it, but it is hard to believe that within a few years, even within one generation, behaviour will change very much unless additional factors, such as media campaigns or the appearance of an unexpected price advantage, trigger a change.

If these rather sceptical ideas about the social acceptance of e-commerce in the coming years hold, the market share of e-commerce will grow not nearly as fast as was assumed two years ago (see 'Survey on E-Commerce', *The Economist*, 26 Feb., 2002: 2). In consequence, if shopping is responsible for 12 per cent of all passenger transport in Germany, the impact on transport demand of e-commerce will be nearly negligible. Even if e-commerce were to attain a market share of 20 per cent, only 2.4 per cent of all passenger transport could be saved. If one adds the additional freight transport, the overall saving potential is even smaller or might vanish completely.

Digitalisation of products

One of the crucial features of the Internet is its ability to replace materialised products by digital ones and distribute them instantanously at almost no cost to millions of customers or users. So, at least in theory, the Internet has the potential to replace materialised products, or services with information content, by new digital ones: in the form of CDs, newspapers, magazines, books (if not for the shelf at home), banking services etc. No wonder that all these products are offered, more or less successfully, via the Internet today. Music is in the lead, but ironically not legally. With the option to download and illegally burn CDs at home, music CD sales fell by more than 10 per cent in 2001. Up to now all efforts to stop this trend by the music industry have failed despite the fact that Napster, the most prominent supplier of recent years, has been forced to terminate these services to its mainly young customers. However, the changes in music markets to overall freight transport are negligible. Even if all music and movies were sold and distributed over the Net, it would change freight transport demand by less than one per cent. The impact on passenger transport would hardly be much larger.

The product with the greatest potential to reduce transport demand in Germany is newspapers. Nearly one in two German households receives a daily newspaper. Replacing those newspapers by digital editions would save millions of tons of paper a year and much transportation of them. However, up to now there are no signs of such a substitution. It seems that, as long as cheap digital paper is unavailable, there will be no reduction in demand because of the Internet. However, the volume of the market for newspapers and magazines on the Internet depends not only on the existence of digital paper. More alarming for newspaper companies are shifts in the advertising market. The bulk of the costs of a newspaper are born by advertisements. If the Internet cannibalises those markets either by regional exchanges for cars, housing and jobs, or via direct mailing, newspapers would be pressed to increase prices significantly, which could then endanger their survival.

In the banking sector e-banking has reached a certain market share. This growth is not mainly demand-driven by Internet minded people but rather pushed by banks to save costs in mass transactions in over-the-counter business. Here the main impact on transport demand lies in passenger transport whereas freight transport is nearly unaffected.

It can be concluded that the shift to trade in some information products and services online leads to little or no reduction in overall freight transport demand.

Example: transport demand and resource intensity of B2C in books

Traditionally, books (in Germany) are distributed through a few large wholesale traders, each supplying around 8 million people through the bookstores. These wholesalers have their own large warehouses organised by highly sophisticated logistics. Books are distributed daily to the shops. To complete the distribution chain, the final trips of customers have to be added. Such a delivery system has to be compared with that of an electronic bookseller, which uses postal parcel services such as Deutsche Post AG, DPS (Deutscher Paket Service), or UPS (United Parcel Service), and which have just one central warehouse for the whole country.

Thus, it is not the book itself (whether one day it might be delivered online or in paper copy) but the transport from the wholesaler to the final consumers' home that is important. Thus it will be assumed that the production itself and the structure of distribution from the printing companies to the few wholesalers remains unchanged.

The first difficulty in such an analysis is the selection of typical services out of the millions of alternatives which can be compared. Results can vary significantly depending on the assignment of transport, inputs or emissions, to shared services such as the distribution of items by a parcel service or shopping during the daily trip to work. Here allocation in general is based on gross weight of the different services. Thus, road transport via truck and car are added by using the gross weight of the vehicles. However, in some cases estimated monetary values are used as the indicator in assigning transport services. Visiting a bookstore on the way to work is only an additional service. On a shopping trip, allocation on the basis of money seems to be more appropriate than weight as, otherwise, daily food articles (e.g. milk) would completely dominate the picture. Nevertheless 2.4 vehicle-kilometres as an average trip to visit a bookstore has been calculated by integrating estimates for the different transport modes (cars, public transport, walking, work, leisure, shopping), average distances and combined services for clients of bookstores. Because of the huge number of parameters, the range is 1.2 to 3.8 vehicle-kilometres. Nevertheless, such figures depend also on the specific German environment and might be completely different in other countries.

People sometimes buy more than one book at once. On the other hand, some people visit a bookstore just to look around. And a certain number of people place an order at the first visit and take the book home on the second visit. Therefore here it is assumed that both effects cancel each other out and that consumers just take away one book on each trip/visit. Table 4.1 shows that the few kilometres of final delivery – whether via parcel service or on behalf on the consumer – are of huge importance for the final impact on gross transport demand.

Table 4.1 B2C best estimates of material/energy flows for book retailing

	Transport demand gross tons/km per book		CO_2 grammes per book		Material input grammes per book[a]	
	Amazon	Traditional bookstore	Amazon	Traditional bookstore	Amazon	Traditional bookstore
Supplier transport	0.63 (0.31–1.34)	0.18 (0.12–0.24)	280 (160–620)	7 (3–13)	1080 (680–2150)	34 (18–57)
Consumer transport	–	3.29 (2.22–6.13)	150 (90–240)	650 (510–1060)	420 (250–690)	2190 (1720–3570)
Deposit + store			41 (38–44)	530 (500–560)	230 (180–270)	3020 (2350–3700)
e-shopping (PC)			140 (40–270)		910 (290–3300)	
Space for PC (5 m^2)			400 (260–580)		500 (400–680)	
Total	**0.63 (0.31–1.34)**	**3.48 (2.34–6.37)**	**1000 (600–1950)**	**1200 (1000–1700)**	**3150 (1800–7100)**	**5200 (4100–7300)**

Note

[a] For example the 5200 grammes per book delivery for the traditional bookstore includes the weight of the fuel for heating the store, the weight of carbon for producing the electricity for the store, the weight of the overburden and iron ore for the car used (a proportional share of course), the share of gravel for the construction of the roads used for taking the book home and all other such material inputs.

Source: Author's calculations.

Obviously in the case of buying a book via B2C long distance transport is much larger compared to traditional distribution structures. One online-retailer has fewer clients, thus economics of scale call for a more centralized distribution centre which enlarges transport distances. Nevertheless, the distance-efficiency of parcel services is higher than the wholesalers' daily delivery system, reducing the difference when gross transport demand is compared. However, what is decisive for the results is the advantage of the final delivery via parcel services, compared to conventional shopping trips. Overall, the transport data show that, in this example, B2C offers a significant potential reduction of transport demand.

For the assessment of the impact on sustainability of online retailing of books, it is not sufficient to consider just transport logistics. In the long run, B2C will have a dramatic impact on the number of retail stores itself. Thus, the inputs for maintaining a traditional store and the warehouse facilities of online retailers have to be included in the analysis. Amazon, the most prominent e-bookstore, is reported to require only one-thirteenth of the store-space of a conventional shop (Grießhammer and Strigl, 2000). Additionally, in the calculation of the resource intensity of e-commerce, the hardware and the electricity for online-shopping have to be accounted for (see Stiller, 2002). Finally, as long as a separate PC is the most common device to go online at home, space for the computer and desk must be included in the analysis. In the calculation, 5 square metres have been included for this purpose and assuming a PC service usage equivalent to buying 3 books per day.[4] Calculation of the material input required for online shopping and the CO_2 emissions yield some surprising results.

First, it appears that long-distance transport contributes only negligible quantities to the overall resource intensity. Resource consumption for transport is clearly dominated by the final distribution. Results depend on the average distance the vans of a parcel service have to travel for the delivery of just one (or a few) book(s) to the final consumer, whether the distance is about 250 m, 400 m or even 1,000 m, and how efficiently people organise their own shopping behaviour for books, whether they combine it with going to work, errands, or other shopping. Integrating over a huge number of different transport modes, allocation of cases and types of shopping behaviour show that there exists a certain advantage for buying the book over the Internet. Nevertheless, ranges in both cases overlap as results depend on the specific circumstances of each case. The advantage in efficiency of parcel delivery compared to a single trip to a bookstore vanishes if such a trip is combined with other purchases or the journey to work. Also, the figures presented here assume that about one in four books is taken home by public transport, bicycle, or even walking. However, parcel delivery should become more efficient once e-commerce becomes more common due to smaller travel distances from customer to customer, resulting in a further efficiency advantage of B2C in transport.

Secondly, data show that inputs for stores are far from negligible. In fact, their effects dominate the resource intensity of long-distance supplier transport by far. If B2C is more resource efficient than traditional retailing, it is because of the smaller shopping space provided and not because of transport efficiency. For example the average electricity consumption per book sold in large retail stores is about 0.5 kilowatt-hours according to Mayersche Buchhandlung (1997).

Thirdly, the inputs that actually have to be assumed to be related to online-shopping via PC are significant but of a smaller size than those for traditional shopping. Data in the calculation are based on 15 minutes online time per book bought. However, as these figures do not rely on any user statistics, real online-time might be significantly shorter, or even longer if people tend to surf around a homepage before buying. Moreover, there are only best estimates available for the material flows per hour of surfing. About 15 per cent of the CO_2 emissions, and an even higher percentage of all resources required for an online purchase of a book, are related to the production of the electronic equipment.

Besides these technical uncertainties, user behaviour has an enormous impact on the final results. The total number of services which a desktop PC provides over its lifetime must be considered. For private use these might range from a few hundred to some 10,000 services. Here inputs are assigned according to user-time. Thus, e-shopping of somebody playing computer games all day would definitely consume a lower amount of resources than shopping on a computer using only a few minutes per day. Moreover, often desktop PCs tend to be kept running, even when not providing services. Those inputs have to be allocated to the services provided.

Overall, ranges presented here should be regarded as best estimates, leaving plenty of room for adjustments if better data become available. In the long run, improved hardware efficiency (flat screen etc.) and higher user-frequency might lead to reduced inputs per service. However, changes in consumer behaviour, e.g. by striving for permanent access, might counteract those developments.

Finally, results show the importance of even more indirect effects. In this example additional space requirements of five square metres have been assumed yeilding a significant contribution to the total results which dominate the CO_2 emissions in this example. Nobody knows what impact the electronic revolution will have on citizens' demand for living space. It might be that people will not spend more time at home, but will either shop elsewhere via mobile phones or will use the time for other outdoor activities. But it may also happen that additional activities undertaken at home will increase the demand for more luxury housing, inducing huge new material flows.

Conclusions

At first glance the book example supports an optimistic view relating to the impact of the Internet on transport demand whereas the impact on broader sustainability indicators such as resource consumption and CO_2 is more ambitious. However, ranges indicate that total impact on transport demand depends very much on the underlying assumptions about the final delivery of goods so that the future impact of B2C can hardly be predicted. In the example presented here the resource efficiency of electronic shopping for books would increase if the hardware were included in other equipment like TV sets. Moreover, user patterns in B2C are far from being known. Most of all, in the example the good itself is not dematerialised.

Indeed, the book-example would look quite different if the next step in the information society is considered. Namely once more consumer friendly interfaces for electronic consumption exist, the need for delivering a hardcopy might phase out. In this case, no transport and warehouse is required, only a re-usable interface to show the content. However, the book example also indicates the limits due to our culture. How can one wrap up an electronic book as a present or put it in the living room on the book shelf? One can imagine that one day both these behaviours will be replaced by other habits, but this will emerge much slower than the development of IT-technology will allow.

Moreover, one has to remember the total volume of the potential virtual products. Food, housing and travel, which are responsible for the bulk of consumers' resource consumption, allow efficiency gains in their production and supply chains, but can never be virtualised. As productivity growth through the Internet is high only in markets with a large number of customers or transactions and digitalizable products (see for example Simon, 2002), e-commerce does not affect many industrial and retail markets, or only to a small degree. Therefore, the direct impact of e-commerce on transport demand will be much smaller than often expected.

Finally, the example shows the importance of secondary effects. There is no hard statistical evidence on a macro-economic level that the Internet has a smooth growth effect on (German) freight and passenger transport demand. Indeed, deregulation and reduction in transaction costs by online booking especially in air transport by low-cost-carriers have triggered a further increase in this environmentally problematic transport mode – until 11 September 2001.[5] Thus, rebound effects obviously impede the Internet's ability to deploy its enormous potential for reducing transport demand, technology for which exists without a doubt. It will not free us from the task of changing transport demand so that it is met in a more sustainable manner.

Notes

1. The idea of eco-efficiency and dematerialization as a proxy for sustainable development means that the ecological need for a significant absolute reduction in

resource consumption and emissions can be achieved by radical new technologies which even allow an *increase* in the overall level of *services* consumed. Using advanced environmentally sound technologies seems to be far more attractive than reducing consumption. There are plenty of examples showing that a factor 4 improvement in eco-efficiency over the entire life-cycle of products and services can already be achieved with existing technologies (von Weizsäcker *et al.*, 1996).

2. The inventory to sales ratio has been declining in the USA in the second half of the 1990s. Although the boom in this period has been partly responsible for this phenomen, levels are lower than during other boom periods. It seems that with the emergence of the Internet inventories have been declining.
3. The emergence of 'food taxis' in Europe seems to indicate the opposite. However, in this case there is no cheaper alternative as delivery is free if the order exceeds a minimum volume.
4. To calculate material intensity or its inverse, resource productivity, the energy (weight of the fuels used plus all related material flows for the extraction and refining of the fuel or the production of electricity), plus the material inputs for the production of a PC are summed. In this case most of the input will be energy for the production of the mother board and the screen (the other material input for the single parts – e.g. sand for chips – causes relatively small material flows). This total (here the energy total) is set off against the total amount of services delivered by a PC during its lifetime. The latter is more difficult to calculate. In fact a simple unquestionable indicator to account for the services provided by a PC does not exist. How can we sum up playing a game for several hours, writing an email, checking the bank account, and buying a book? For the purposes of the calculation here 'user time' is used as an appropriate indicator: an average user time of about one hour per day (that does not include the time the PC is running but nobody is using it). This appears to be a good guess for online clients of Amazon. Of course, many youngsters use their PCs much more than one hour per day (e.g. to download music) but they do not often buy books online and they do not represent the average online client. For our purposes the median hours of PC use per person is decisive, not the average hours.
5. Ryanair and EasyJet are reported to sell 90 per cent of their seats on the Internet. See: 'Flag carriers at half-mast' *The Economist*, 2–8 March 2002: 61–2.

5
Automotive Pollution Control Technologies

Carlos Destefani and Elias Siores

Introduction

Since the beginning of the industrial revolution in the eighteenth century, factories have been throwing waste products in solid, liquid, and gaseous states into the water and air. Waste was disposed of in the most convenient way technology could offer at the time. With the introduction of automobiles the standard of living for many people has improved through the increased capacity for mobility, but another source of gaseous pollutants was introduced to our planet. The development of industries and vehicles burning fossil fuels as a means of using stored energy for the purpose of mobility has substantially contributed to the increase of gaseous pollutants released to the atmosphere. Attempts to reduce these effluents by applying emission treatment devices have been made throughout the years. Since the middle of the last century the output of waste has increased to the point where its impact can no longer be ignored, and environment protection has become a general concern for humankind.

Emission regulation standards have set new levels of emissions to be achieved within the next few years in order to minimise atmospheric pollution. The European Union will introduce the Euro IV standard in 2005 and the US Standard, Tier 2, will be implemented by the beginning of 2004. The new emission levels imposed by these standards clearly indicate the need for development of a new generation of technologies. Even though we do not claim that sustainable transport will be brought about by new technology alone, without question technological improvements in vehicle engines will form an important part of the solution.

In this chapter we first briefly discuss the gaseous pollutants from combustion of fossil fuels, and we then review the forms that the new generation of vehicle technologies, aiming at reducing pollution, will take. We consider current research and development of electric vehicles, fuel cells and hybrid electric engines that will greatly reduce both fuel consumption and emissions. We point out, however, that petrol/gasoline engines will still be with us for many years to come, and we discuss current research into reducing pollution from such engines. Finally we report a new development in pollution control that uses microwave energy to treat exhaust gases from standard internal combustion engines, a development capable of extracting most of the carbon from engine emissions along with most other pollutants.

Pollutants from vehicle engines

Combustion of petrol/gasoline and diesel in internal combustion engines results in the following emissions: carbon monoxide and dioxide (CO, CO_2), hydrocarbons (HC), nitric oxide (NO), nitrous oxide (N_2O), oxygen (O_2), water (H_2O), and some sulphur oxides present in the exhaust gases from fuels containing sulphur. In the combustion products, the first five components and sulphur oxides are considered pollutants. Carbon, sulphur and nitrogen oxides pollute the atmosphere, accelerate the greenhouse effect and contribute to acid rain.

The major gases in the atmosphere, nitrogen and oxygen, are transparent to both the radiation incoming from the sun and the radiation outgoing from the Earth, so they have little or no effect on the greenhouse effect (atmospheric warming). Gases not transparent to incoming and outgoing solar radiation (greenhouse gases) are water vapour, ozone (O_3), carbon dioxide (CO_2), methane (CH_4), nitrous oxide (N_2O), and the chlorofluorocarbons (CFCs). In automotive exhaust gases, CO_2, NO, N_2O and HC are present and released to the atmosphere. Controlling CO_2 production from combustion processes is becoming one of the major pollution concerns of industrialised countries.

Carbon monoxide (CO) is a pollutant that is readily absorbed in the body and can impair the oxygen-carrying capacity of haemoglobin. Impairment of the body's haemoglobin results in reduced oxygen to the brain, heart, and tissues. Short term over-exposure to CO can be critical to people with heart and lung diseases, capable of causing headaches and dizziness. During combustion, carbon in the fuel oxidises to form carbon dioxide (CO_2). However, complete conversion of carbon to CO_2 is rarely achieved in practice and some carbon oxidizes as CO.

Nitric oxides (NO_x) are pollutants mainly produced by the combustion processes. In industrial nations over 50 per cent of NO_x results from road traffic (Pott, A. *et. al.*, 1998). Molecular nitrogen (N2) accounts for 79 per cent of air and therefore its presence in petrol/gasoline combustion processes results in NO_x; such as NO, N_2O, and NO_2. Controlling their formation is crucial for the generation of NO_x that contributes to the formation of smog and acid rain (Chen *et al.*, 1991). It is estimated that the major source of emissions of NO_x attributable to human activities is from gasoline/petrol combustion in internal combustion engines.

There is increased public awareness of the contribution to global warming caused by vehicle emissions and an ever increasing demand by the public to introduce more environment-friendly automobiles or efficient alternatives in reducing the potential damage caused by motor vehicles, and especially cars, to our environment. Billions of dollars are now being spent by automotive companies and governments on research into and development of alternative sources of energy to power cars instead of petrol/gasoline or diesel fuels. The imperative for such developments comes also from the growing perception of the impending limits to oil production (see Chapter 3). These alternative energy sources have generated the concept of Electric Vehicles (EVs), Fuel Cell Vehicles (FCVs) and Hybrid Electric Vehicles (HEVs).

New vehicle technologies

Electric vehicles

Electric vehicles (EVs) had been introduced in the industrial nations by the end of the nineteenth century. They have since attracted interest from time to time when environmental or energy problems came into consideration. The performance of EVs is still in the process of technical improvement: in batteries, motors, control systems and other components. The cost and performance of these components has not reached an adequate level for their introduction into the market with a significant effect on the environment or energy consumption (International Energy Agency, 1993).

EVs driven by an electric motor powered by batteries present the environmental benefit of a zero-emission vehicle, provided the electricity is generated at source without the use of fossil fuels. The environmental benefits of this alternative are significant but there is a lack of developed technology associated with batteries providing for normal driving. Batteries are the most clean alternative energy source for vehicles, but the technology involved in their development has proved to be problematic. Currently available batteries are still not capable of providing performance acceptable to the consumer. Their limited range and slow battery charging process are constraints that still need to be overcome. General Motors' EV-1 and Honda's EV-Plus models have these difficulties. However, American, Japanese and European companies are actively allocating resources and investing in the development of advanced batteries that could power EVs for everyday use.

The most important factor necessary to make these vehicles available to consumers is the development of batteries both at a lower cost and with a performance comparable with conventional vehicles. Batteries in current use provide only a limited travelling range before they have to be recharged, and this process takes several hours. EVs are more efficient in the use of energy than vehicles with conventional combustion engines, but they are less efficient at *storing* energy. The limited driving range of EVs is the most difficult problem to overcome and the greatest disincentive to their widespread production. All these problems (driving range, battery charging time and costs) present significant difficulties, and solutions should be implemented in combination (International Energy Agency, 1993). Possible future solutions to the problem of limited range are (a) the development of a high specific energy battery; and (b) reduced energy consumption by the vehicle. Solutions to slow energy replenishment are: (a) rapid battery charging facilities; (b) battery replacement; and (c) aluminium/air batteries, and fuel cells (see discussion below).

High specific energy is required, that is high power output in relation to the battery's weight (watt-hours/kg). Besides this, a suitable battery for EVs also requires:

- long cycle life;
- low cost;
- safety;
- simple maintenance;
- the ability to be recycled;
- no risk of causing environmental pollution when discharged;
- the ability to be recharged rapidly.

Since the 1920s, the popular choice of power source for EVs was the lead/acid battery. These batteries can vary widely in design, choice of component materials and details of manufacture. The following advantages have been found when compared to other kinds of batteries: (i) the lead acid battery has the highest cell voltage of all battery systems using an aqueous electrolyte; (ii) it has the ability to supply both high and low currents over a comparatively wide range of temperatures; (iii) it has a high degree of reversibility with a satisfactory energy efficiency; (iv) it has an acceptable level of charge retention during storage or inoperative periods in a charged condition; (v) it can be manufactured at low cost compared with other batteries; and (vi) there already exist established manufacturing and recycling plants (Rand *et al.*, 1998). For EV applications, the great disadvantage lies in the heavy weight of the basic material (lead), a relatively short cycle life and long recharging time. In fact only a small fraction of the lead is active in the battery. Environmentally safe disposal of the lead can also present serious problems.

Research into and development of other types of battery, based on alkaline electrolytes have been conducted since the late 1890s. In alkaline batteries, the electrical energy is derived from the reaction of a metal with oxygen, which could be used either on its own as in metal/air systems or be held in the form of metal oxides adding weight to the battery and reducing the specific energy. On an energy-to-weight basis, the best negative active material is hydrogen. The nickel/gaseous-hydrogen system displays a cycle life excellent for EV applications. Nevertheless, there remain the cost and safety problems associated with storing hydrogen gas. Recent improvements in alloy technology, however, have allowed the nickel/metal hydrogen system to become a promising candidate for EVs. Alkaline batteries include: nickel/iron, nickel/cadmium, nickel/zinc, manganese/zinc, nickel/metal hydride, aluminium/air, and zinc/air batteries (Rand *et al.*, 1998).

The nickel oxide electrode in alkaline batteries may be replaced either by an air electrode or a silver oxide electrode. Silver/zinc and silver/cadmium batteries are known for applications in the military defence field. The nickel/zinc battery is the most attractive because of its relatively high specific energy and lower cost compared with cadmium. Nickel/metal hydride batteries are in constant development. These batteries provide high specific power, making them suitable for hybrid car applications (see below).

The future of metal/air batteries lies in the future development of the air electrode. Zinc/air systems are being developed in Israel and by an Australian–US consortium, Power Air Tech Pty Ltd, formed in September 1997 to consolidate and develop further the technologies created in the Lawrence Livermore National and Lawrence Berkeley Laboratories. In the long term, an attractive alternative for EVs is aluminium/air batteries. At present, limitations of power and electrical efficiency need to be overcome before such batteries are considered a real option as an energy source for EVs. On the other hand, lead-acid batteries have over the years demonstrated a satisfactory level of power density and cost. Their low price has made them more popular than any other type described above.

In parallel to battery requirements listed previously, research on weight reduction in cars promises further improvements. In a typical EV the weight of the batteries is in the order of 500 kilograms, half the weight of a conventional small car. In Australia, the Commonwealth Scientific and Industrial Research Organisation (CSIRO) has developed a power pack made up of batteries and supercapacitors weighing just 110 kilograms (Axcess Australia, 2001). The principal attributes in EVs batteries are summarised in the following Table 5.1.

While the development of batteries to power EVs are in continuous flux, aiming at optimising their efficiency and cost, automotive manufacturers have recently released new EV models. Japan's major auto manufacturers unveiled the state-of-the-art EVs that are very different in performance, vehicle type, and cost from their predecessors. As reported by the Japan Electric Vehicle Association (JEVA, http://jeva.or.jp), these new EVs were developed in line with the California Zero Emission Vehicle mandate introduced in both California and Japan since 1997.

In September 1996, Toyota Motor Corporation started to market its production electric passenger car RAV4 EV, the first commercial electric vehicle powered by nickel-metal hydride batteries that provide 215 kilometres travel distance per charge and a top speed of 125 kilometres per hour. In the development of these batteries, Panasonic EV Energy joined Matsushita Electric Industrial Co., Matsushita Battery Industrial Co., and Toyoda Automatic Loom Works. Toyota has estimated annual sales of approximately 100 vehicles initially to municipalities and utilities. At that time, Toyota planned to

Table 5.1 Principal attributes of batteries for electric vehicles

Battery	Principal attractions	Limiting factors
Lead/acid	• Established industry. • Lowest cost battery. • Sealed (maintenance free).	• Low specific energy. • Moderate cycle life (~500). • Poor low-temperature performance.
Nickel/iron	• Established, robust battery. • Long cycle life.	• Moderate specific energy. • High maintenance. • High cost.
Nickel/cadmium	• Established industry. • Long cycle life.	• Moderate specific energy. • Poor performance above 35 °C. • Highly toxic component. • High cost.
Nickel/zinc	• Reasonable specific energy. • Good all-round candidate battery.	* Short cycle life (~300).
Nickel/metal hydride	• Reasonable specific energy. • High power. • Long cycle life. • Tolerance to overcharge and overdischarge. • Sealed (maintenance free).	• High self-discharge. • Poor charge-acceptance at high ambient temperature. • High cost.
Metal/air	• Long EV range before re-plating or mechanical recharging.	• Complexity. • High volume (low energy density). • Low overall electrical efficiency. • Re-plating technology not yet established.
Zinc/bromine	• Reasonable specific energy. • Demonstrated in EVs. • Low-cost components.	• Low power output. • Shunt-current problems. • Safety concerns (Br_2 release).
Sodium/sulfur	• High specific energy. • Low-cost electrode materials. • No self-discharge.	• High operating temperature. • Corrosion of cell components. • No overcharge capability. • Safety concern (fire).
Sodium/nickel	• High specific energy. • Discharge-state assembly. • Long cycle life. • No self-discharge. • Successful fleet trials.	• High operating temperatures. • Low specific power at 80%. • Cost?
Lithium-ion	• High specific energy and power. • Long cycle life.	• Early stage of development. • Accurate control of charge voltage necessary. • Safety? • High cost.
Lithium/iron	• Reasonable specific energy. • Good power output. • Long cycle life.	• High operating temperature. • Only a bipolar battery will meet specifications for Evs. • No pilot production despite 20+ years of research.

Source: Rand *et al.* (1998), p. 486.

produce about 320 RAV4 EVs for the California market by the year 2000. Information on this EV is available in http://www.toyota.com (2001).

Nissan Motor's Prairie Joy EV is the world's first production EV powered by lithium-ion batteries. These batteries were jointly developed with Sony Corporation, providing the EVs with a range of more than 200 kilometres per charge and a 120 kilometres per hour top speed. In 1997, Nissan planned initial sales of 30 Prairie Joy EVs to Japan's municipalities and corporate users. Nissan also unveiled production of the Altra EV (R'nessa in Japan). The Altra EV is a four-passenger compact van powered by lithium-ion batteries to be produced for the California market as well as for Japan (JEVA).

Honda EV Plus was designed and manufactured by Honda Motor Company in April 1997. The vehicle is powered by a nickel-metal hydride battery with a driving range of 210 kilometres per charge and a top speed of 130 kilometres per hour (JEVA).

Fuel cells

In 1839 Sir William Grove discovered that electrochemical energy is converted into electricity and heat when combining hydrogen and oxygen, and he invented a process to make this happen. This process originated the Fuel Cell (FC). Since then, as with batteries, several types were developed in the nineteenth and twentieth centuries. NASA scientists utilized this discovery in the 1950s to develop fuel cells to power space exploration vehicles (US Fuel Cell Council, 2001 http://www.usfcc.com/internet.htm). Over the years, many classifications of fuel cells have appeared, due to the number of variables amongst FC systems, such as types of fuel and electrolyte, operating temperature, primary and regenerative systems, and direct or indirect systems (Blomen *et al.*, 1993). FCs can be classified by the type of electrolyte used: alkaline (normally potassium hydroxide, KOH), acid (mainly phosphoric), molten carbonate (the electrolyte is 62 per cent Li_2CO_3 and 38 per cent K_2CO_3), and solid oxide and solid polymer (proton-conducting membranes, now mainly DuPont's Nafion and Dow's membrane).

Since the late 1950s, FCs have been under rapid development for implementation in power plants, space and terrestrial applications. In the last application, FCs have to reach a state in which they prove to be technically and economically better than conventional power plants. The relative advantages and disadvantages of different FCs are assessed on a quantitative basis, as follows:

(i) fuel efficiency;
(ii) power density, which is important from the point of view of minimising the weight, volume, and capital cost of the power plant;
(iii) projected rated power level;
(iv) projected lifetime;
(v) projected capital cost.

In automotive application, FCs can be used to power buses and cars. There are different types of FCs, but the most common one found is the Proton Exchange Membrane (PEM). FC systems have three sub-systems: the fuel processor, the fuel cell stack, and the power conditioner. Several more components such as pumps, compressors, heat exchangers, motors, controllers and batteries are utilised in the system. These components require special design and manufacture. The fuel cell stack is the central part of the fuel cell engine and consists of hundreds of individual PEM fuel cells assembled to produce enough electricity and direct current (DC) from the chemical reaction of hydrogen and oxygen to power a car or bus. There are fuel cell stacks with electric capacities from 50 to 70 kilowatts and higher, powering prototype passenger cars, minivans, Sport Utility Vehicles, and buses. In recent years, great improvements in stack efficiency, weight and size reduction have been achieved, but more remains to be done.

FCs are DC power generators, and in some applications DC is converted into alternate current (AC) to run AC induction motors by means of AC motor controllers, whereas in other cases DC motors are used. Much of the research and development and resources allocated to EV drive trains in the last few decades have been applied to Fuel Cell Vehicles (FCVs) (US Fuel Cell Council, 2001).

These new EVs have an electric drive train powered by either a battery or a fuel cell – both energy sources converting chemical energy into electricity. The reactants in a battery are stored internally and the batteries need to be recharged or replaced when the energy is used. In a fuel cell powered EV, the fuel is stored externally in the vehicles' fuel tank and air is obtained from the atmosphere. As long as there is fuel in the vehicles' tank, the fuel cell will produce electricity and heat. Fuel cells, when compared to batteries, are smaller in size, lighter, with quick refuelling and longer range (Thomas *et al.*, 2001).

Fuel cell technology has proven itself to be a greener alternative to IC engines and now every automotive company in the world is developing them. As mentioned previously the pure hydrogen cell has some advantages over others. Some automotive companies are trying to find gasoline-like hydrocarbon fuel cells because they are more advanced technically and have most potential for use in commercial vehicles. These are the Methanol Fuel Cell Vehicles (MFCVs). They offer environmental benefits while retaining the performance and range of today's IC engines. Methanol emerges as an ideal hydrogen carrier for vehicles because it is liquid at room temperature and ambient pressure (American Methanol Institute, 2001). It is a simple molecule consisting of a single carbon atom linked to three hydrogen atoms and one oxygen–hydrogen bond (CH_3OH), produced from natural gas or distilled from coal. Building a methanol infrastructure to service such vehicles, similar to the LPG (liquid petroleum gas) infrastructure introduced in some countries such as Australia, would not be difficult.

Daimler-Benz was the first company compiling emission data from MFCs. They found zero emissions for NO and CO and extremely low but significant levels of HC, similar to a direct injection diesel engine in terms of CO_2 – indicating no improvement in greenhouse-gas emissions (Motavalli, 2000). Table 5.2 illustrates the prototype fuel cell vehicles introduced to date by major automotive manufacturers.

The principal obstacle to the widespread introduction of FCVs is the establishment of a refuelling infrastructure. If fuel cell vehicles can be designed

Table 5.2 Prototype fuel cell vehicles

Manufacturer/vehicle type	Year shown	Fuel type
BMW		
Series 7 Sedab	In development	Hydrogen
DAIMLERCHRYSLER		
NECAR (van)	1993	Gaseous hydrogen
NECAR 2 (mini-van)	1995	Gaseous hydrogen
NECAR 3 (A-class)	1997	Liquid methanol
NECAR 4 (A-class)	1999	Liquid hydrogen
Jeep Commander 2 (SUV)	2000	Methanol
NECAR 5 (A-class)	2000	Liquid methanol
Ford Motor Company		
P2000 HFC (Sedan)	1999	Hydrogen
THINK FC5	2000	Methanol
General Motors/Opel		
Zafira (mini-van)	1998	Methanol
Precept	2000	Hydrogen
Honda		
FCX-V1	1999	Hydrogen
FCX-V2	1999	Methanol
Mazda		
Demio (compact car)	1997	Hydrogen (stored in a metal hydride)
Nissan		
R'nessa (SUV)	1999	Methanol
Renault		
FEVER (station wagon)	1997	Liquid hydrogen
Laguna Estate	1998	Liquid hydrogen
Toyota		
RAV 4 FCEV (SUV)	1996	Hydrogen (stored in a metal hydride)
RAV 4 FECV (SUV)	1997	Methanol

Source: American Methanol Institute (2001), p. 13.

to run on gasoline and diesel fuels it would overcome this problem. The major research and development efforts in this area are directed towards the Gasoline Fuel Cell Vehicles (GFCVs) and Diesel-Battery Hybrid Vehicles (GBHVs).

The GFCVs are a decade behind other technologies in terms of development and several years behind in terms of commercialisation effort compared to MFCVs. Today's gasoline has components presenting challenges that need to be overcome in refineries; such as sulphur levels. This technology is under development, but at the same time methanol has a number of reasons for being the preferred fuel choice for fuel cells (American Methanol Institute, 2001):

- Gasoline is not practical to use in a direct PEM fuel cell, locking the industry into continual use of reformers.
- Gasoline reformer development presents a high level of complexity that will delay the commercialization of FCVs.

Major automakers, and Ford in particular (Birch, 2001), are working vigorously to achieve a FCV and overcome the related manufacturing and cost problems. But at the same time a cultural change is required that will support an infrastructure to fuel FCVs. The Focus FCV is the latest model from Ford using a PEM fuel-cell stack. It is 50 per cent smaller and delivers 30 per cent more power output than previous models.

Hybrid electric vehicles

Another technological development that responds to present battery limitations in EVs is the concept of Hybrid Electric Vehicles (HEVs). These vehicles have both an internal combustion (IC) engine and a battery-powered electric motor working together to drive the car. HEVs reduce smog-forming pollutants and cut greenhouse emissions between a third and a half. Further research on more efficient cars will bring even lower emission levels. HEVs sacrifice the zero-emission environmental benefit of EVs, carrying a small IC engine to gain the range and quick refuelling capacity of a petrol/gasoline vehicle. HEVs have been conceived as an alternative to overcome the shortfall in battery technology. In HEVs an on-board generator powered by an IC engine is used for longer trips (Prange, 2001).

HEVs have a number of potential advantages over conventional IC vehicles and EVs:

- HEVs generate their own electrical power addressing the two main problems of EVs, limited range and extended recharging time.
- At cruising speed, matching the generator power and motor capacity with vehicle power demand will provide a range limited to the fuel tank capacity, if the balance is maintained.
- In size, the battery pack required by HEVs is smaller than EVs because an IC engine is incorporated in the propulsion system. Nevertheless, the overall weight and dimension of the propulsion system is similar.

- When HEVs run at speeds below the cruise speed, when idling or decelerating, the excess energy generated re-charges the batteries.
- The IC engine operates at its optimum efficiency irrespective of road speed.

The HEVs have an IC engine and electric motor as the source of mechanical power in different configurations. Based on the criterion of a propulsion torque source, if the electric motor is the only source of propulsion, the configuration is called 'series'. If both – when the IC engine and electric motor are utilised to create the propulsion torque – the configuration is called 'parallel'.

In a series configuration, the electric motor is the only traction power source. Therefore, it needs an energy converter from mechanical (IC engine) to electrical and this system comprises the IC engine, generator, electric motor and battery where the first three are connected in series. The battery and generator for electric energy supply are in parallel. The IC engine capacity is designed to provide energy continuously at maximum cruising speed. This means that the vehicle range depends on the fuel tank capacity. This configuration has the possibility of operating the IC engine at constant speed to minimize emissions and maximize efficiency. On the other hand, the series configuration has high costs associated with the weight and propulsion system. The three energy conversion stages implicate energy losses and consequently reduce the overall efficiency of the system.

In a parallel hybrid, the electric motor and IC engine are both utilised as the propulsion power source. One of them is used at steady power level and the second source is on stand-by to supplement power in peak consumption circumstances. Electric motors provide more flexibility and in most cases are used to supplement energy. This configuration presents lower weight and cost; the generator is not required in the system and therefore the overall system efficiency is improved. The disadvantages of the parallel configuration are based on the complex transmission that is necessary to combine both torques. A smaller electric motor is a disadvantage in regenerative braking (a process converting kinetic energy into electricity when the HEV is braking or driven downhill) and requires a complex control system to distribute the torques (ORTECH International, 1993).

General Motors Corporation has worked on the development of the Gen2 Stirling Series since 1993. This car has a stirling engine as a power unit. This regenerative heat engine uses an external combustion process with a closed-cycle, powered by the expansion of a gas when heated and followed by the compression of the gas when cooled. The project, since early 1999, proved the technical feasibility of the series hybrid electric propulsion but failed in meeting commercial viability. Further work is required to address the commercial challenges (Office of Transport Technology, 1999b).

The Ford programme since December 1993, identified two parallel hybrid propulsion system configurations with the promise to achieve low fuel consumption and tailpipe emissions. A lightweight prototype, the P2000, was co-developed with Milford Fabricating Company. The P2000 weighs 2000 pounds (~908 kilograms) and Ford estimated that it will achieve 63 miles per gallon (~25 kilometres per litre). Further work on weight reduction is required since there are 500 pounds (~227 kilograms) of steel and ferrous metal in the vehicle frame. The HEV is equipped with two lead-acid battery configurations or a nickel metal hydride (NiMH) battery. Based on test results the NiMH battery will be implemented in the deliverable vehicle (Office of Transportation Technologies, 1999a). Ford plans to introduce the Escape HEV by 2003. This vehicle offers better fuel economy and lower levels of pollutant emissions than a petrol/gasoline car. The expected range of this HEV is 450–550 miles (720–885 kilometres) compared to 350 miles (560 kilometres) and 100 miles (160 kilometres), for IC engine vehicles (4 cylinders) and EVs respectively (Ford, http://www.hybridford.com/index.asp, 2001).

Japanese automotive manufacturers, Honda and Toyota, in the summer of 2000 introduced two HEVs to the American market, respectively the Insight and Prius. More than 50,000 Prius have been sold already in Japan (Toyota http://prius.toyota.com, 2001)

Petrol/gasoline vehicles

Past and current research and development on EVs, FCs, and HEVs, have been outlined in the previous sections. Technically, these alternatives have proved the viability of their respective technical concepts. However, these alternatives need further development to meet consumer expectations. The conventional IC engine vehicles will be on our roads at least until the infrastructure for the selected option; either EVs, FCVs, or HEVs; is in place.

Over the last few years, automobile manufacturers have attempted to reduce, by various methods, the concentration of dangerous components in the exhaust gas mixture. First, combustion gases were made safer by more environment-friendly fuels, (unleaded petrol) or by improving the air-fuel mixture system in the IC engine before combustion takes place. Secondly, improved safety was achieved by altering the conditions in which the combustion process is taking place within the IC engine. And, thirdly, improvements were achieved by combustion exhaust after-treatment chemical processes – such as catalytic converters installed in the exhaust system removing unburned hydrocarbons and modifying the concentration of CO_2, CO, and NO_x either by reduction or oxidation processes. The catalytic converter is used in modifying the exhaust gas chemical composition and releasing 'better quality' pollutants to the environment, that is pollutants that comply better with the regulations. This technique requires further development to comply with future standards that will become more difficult to meet, with lower emission levels to be achieved within the next few years, Tier 2 in US by the beginning of 2004 and Euro IV by 2005 in Europe.

Environmental and political issues came together in the development of the catalytic converter. In 1981 stricter emissions regulations were established in US controlling NO_x. This requirement set new conditions for catalytic converters. Initially, catalytic converters were used to control hydrocarbons and CO, but then the three-way catalysts appeared. In this catalytic converter, hydrocarbons and CO are oxidized to CO_2 and H_2O and NO_x is reduced to N_2. Noble metals such as rhodium, platinum and palladium are essential in these converters, all of them containing massive embodied energy from the mining and production processes. The average durability of a catalytic converter is five years (50,000 miles) and the major mechanisms for catalytic converter deterioration are: (a) thermal damage, caused by exposure to high temperatures; (b) poisoning by contaminants in the exhaust; and (c) mechanical failure, for example catalyst support damage (Taylor, 1987). Every technology presents some weaknesses and this particular one does not work in all IC engine conditions. It performs well when the air/fuel ratio is close to the stoichiometric mixture of 14.6 (a quantitative relationship between air and fuel in the combustion process).

Alternative approaches to minimize pollution using electric discharges have been implemented to modify the chemical composition of constituents in the exhaust gas mixture. Reduction of NO_x has been obtained from diesel engine exhaust gas using a superimposed barrier discharge plasma reactor (Urashina *et al.*, 1998). Higashi *et al.*, (1991) reported simultaneous reduction of soot and NO_x in a diesel engine exhaust by discharge plasma. Experiments for NO_x reduction with a N_2 microwave-excited discharge with the addition of NO and N_2 mixture was investigated by Pott *et al.* (1998). Previous work in reducing CO_2 was published by Higashi *et al.* (1985) and Weiss (1985), who have shown that CO_2 concentration in a N_2–CO_2 mixture or pure CO_2 can be reduced by a DC and Pulsed Streamer Corona Discharges respectively. Maezono *et al.*, (1990) have investigated the CO_2 reduction at concentration up to 10,000 ppm using a laboratory scale DC Corona Torch, and conversion of CO_2 into CO in point-to-plate DC corona discharges by Boukhalfa *et al.* (1987). The mechanism of CO_2 reduction in the presence of noble gases has also been studied by Palotai *et al.* (1989) and reductions of up to 40 per cent of CO_2 in Ar–CO_2 mixture gases using a capillary tube plasma reactor were reported.

Experimental rig and findings

This section includes technical details that explain the microwave process capabilities in modifying the exhaust gases' chemical composition. At the Industrial Research Institute Swinburne (IRIS, Australia), work applying microwave energy to treat exhaust gases from IC engines has been undertaken.

The experimental rig for exhaust gas microwave treatment is illustrated in Figure 5.1. It consisted of a 6 cylinder IC engine from which part of the exhaust gases' flow rate (on average 3.5 l/min) was sent, by a hose connected

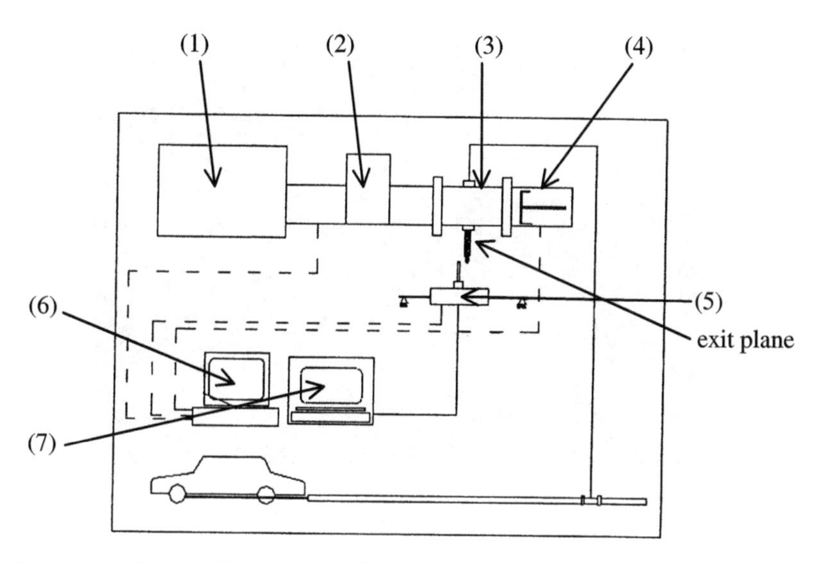

Figure 5.1 Exhaust MIP experimental arrangement

to the exhaust tail pipe, to the plasma applicator, where an exhaust gas Microwave Induced Plasma (MIP), was generated. Chemical composition studies were performed in the exhaust MIP discharge. Microwaves were generated using a Coaxial MG6K microwave generator (1); a rectangular waveguide of 86 mm by 43 mm (width by height respectively) was connected to the magnetron through a high power impedance analyser (2). The exhaust gas MIP discharge was induced in the plasma applicator (3), and a short-circuit type plunger (4) closed the microwave transmission line setting the TE_{101} resonant mode. Quartz glass tubes of 4 mm ID × 6 mm OD were used as discharge tubes. An enthalpy probe (5) controlled by a PC (6) sampled the MIP discharge and sent to an Autodiagnostic gas analyser (7), model ADS9000, before and after the MIP treatment, measuring CO, CO_2, NO, O_2, and hydrocarbons (HC). In Figure 5.2 the ADS9000 gas analyser and the exhaust gas input line as set during the experimental work is depicted.

The concentrations of the different gases in the exhaust gas mixture before and after plasma treatment with the IC engine running at idle speed (750 rpm ± 50) are compared in Table 5.3. The results given are the mean of measurements made at 2 mm intervals, beginning at the exit plane (Figure 5.1) of the discharge tube and extending axially away 10 mm. Results are given for exhaust gases treated and untreated by the catalytic converter and at three different incident microwave power levels and discharging into ambient air. The microwave power absorbed by the discharge is also given.

Table 5.3 shows that the microwave plasma treatment of the exhaust gases greatly modifies their composition. The CO_2 content of the gas mixture is

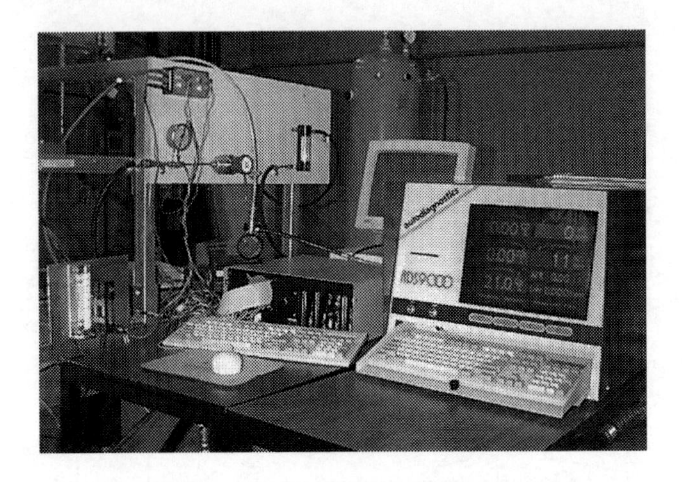

Figure 5.2 ADS9000 gas analyser and exhaust gas inlet line

Table 5.3 Measured concentrations of different pollutant gases in the exhaust gas mixture before and after microwave-plasma treatment

	Exhaust gas							
	Without catalytic converter				**With catalytic converter**			
Incident microwave power (W), [Absorbed]	0	500 [490]	700 [550]	1500 [1450]	0	500 [320]	700 [460]	1500 [490]
CO (%)	0.13	0.04	0.05	0.00	0.00	0.20	0.02	0.15
CO_2 (%)	13	0.5	0.6	0.0	13.7	1.0	0.7	1.0
O_2 (%)	2.5	20.5	19.8	20.9	2.3	19.9	19.7	19.7
HC (ppm)	12	3	2	1	8	1	1	1
NO (ppm)	14	600	600	10	140	620	750	400

substantially reduced at all microwave power levels. This reduction in CO_2 content is accompanied by either a reduction (in the case without catalytic converter treatment), or only a minor increase (in the cases with catalytic converter treatment) in the CO content. The O_2 concentration increases significantly. In low and medium microwave power levels, the NO concentration is increased by the microwave plasma treatment. In the high microwave power level applied, the NO concentration was unaffected in the plasma length studied.

There was evidence that the discharge treatment reduces the hydrocarbon concentrations. Further investigation was undertaken and the hydrocarbon

Table 5.4 Exhaust gas chemical composition: IC engine at idle speed

	CO (%)	CO_2 (%)	O_2 (%)	NO (ppm)	HC (ppm)
Exhaust gas content	0	11.64	4.62	62	8.35
After microwave treatment content	0.095	4.97	12.88	4000	37.62

content was analysed using a gas chromatograph/mass spectrometer. These analyses were conducted to determine and quantify the sub ppm (parts per million) levels of individual hydrocarbons (C1 through C6), n-paraffins and benzene present in levels higher than 100 ppb (parts per billion). It was found that exhaust MIP discharges sampled did not contain any detectable amounts of the targeted hydrocarbon species, confirming the microwave process efficiency in breaking down hydrocarbon bonds too.

Further experiments were conducted in which the final chemical compositions, measured in an exhaust gas environment once the exhaust gas MIP released the energy and returned to room temperature, are summarised in the following Table 5.4. The microwave power applied was 500 W.

Table 5.4 shows the final chemical composition (concentration by volume) experienced by the exhaust gas mixture undergoing the microwave treatment process. The differences reported are the final chemical composition in the exhaust gas released to the atmosphere. The microwave energy reduced CO_2 by 57 per cent, small concentrations of CO and HC were found. Palotai and Chang (1989) reported reductions of up to 40 per cent of CO_2 in Ar–CO_2 mixtures in electric discharges by capillary tube plasma reactor. CO formation by a corona torch plasma device was reported previously by Maezono *et al.* (1990). The free carbon, (generated by the CO_2 dissociation), originated CO and HC in very low levels. Thermal NO was formed by the developed temperatures and measured at high content level. No technique to control NO was implemented in this investigation. Nevertheless, reduction of NO_x by gas discharges has been reported by Urashina *et al.* (1998).

This microwave technology has the potential of greatly decreasing CO_2, one of the most important gases in the global warming effect. The other carbon constituents; CO and HC; do not present the same problem magnitude as CO_2 does. The dissociation levels, between 50 and 60 per cent in CO_2 is a significant achievement to further develop this technology. The carbon generated in the process may be gathered for further processing. Molecular oxygen (O_2) increases its level from an average of 2–2.5 per cent in an exhaust gas before microwave treatment to 12 to 13 per cent after microwave treatment. These dissociation and recombination levels are related to the microwave power implemented in the system.

Conclusions

The current research and development activities in alternative energy sources to power motor vehicles, have been reviewed. HEVs, the latest development, emit lower levels of CO_2 than conventional internal combustion engine vehicles. However, if the electric power generation is achieved with CO_2 production, the problem of contributing to global warming is not solved. The world's major automakers are racing to introduce into the market reliable vehicles, but nobody has succeeded. No clean technology has yet been fully developed. The internal combustion engine has been powering vehicles for over 100 years, and finding an alternative treatment for exhaust gases provides a greater potential application, while at the same time, research into the other technologies can also achieve breakthroughs.

An alternative in treating exhaust gases is by using microwave energy, applicable to cars and stationary sources. This microwave technology has the potential of greatly reducing CO_2, one of the most important gases contributing to global warming. Work optimising the microwave process and controlling NO_x formation will further develop this process to its full potential. Built as an inexpensive device that can be attached to normal exhaust systems, the microwave emissions converter is silent in operation, needs minimum power, does not present any health hazards and can be maintained as part of the normal routine service of an automobile. Constructed of aluminium and weighing only a few kilograms, it will not noticeably increase the overall weight of a vehicle. It could be installed in all normal internal combustion vehicles, old or new. In combination with other technologies, microwave treatment of exhaust gases has considerable potential both for reducing local pollution and cutting greenhouse emissions from transport.

Part II

Regional and National Studies

6

Towards Sustainable Transportation Policy in the United States: A Grassroots Perspective

Cameron Yee

Introduction: policy built for suburban commuters

Transportation policy in the United States has travelled a road of top-down decisionmaking in which federal policy was set without participation from local communities. Government's failure to create a process that addresses community needs through local and regional funding has resulted in a car-dominated transportation system that is not sustainable. A sustainable transportation system should provide people with affordable transportation options to get to work, play, school, shopping, and health care among other transportation needs. At the same time, the system should not have detrimental environmental impacts. However, transportation policies set by government at the state and federal level have focused on the suburban commuter travelling to work during the peak hours at the expense of other users of the system, and the environment. The needs of low income communities, youth and the elderly have often been left out of the policymaking process. This policy framework for transportation has skewed priorities to meet the needs of the middle-class suburban commuter at the expense of urban communities, especially low income communities and communities of color (Yee, 1999).

The focus on peak suburban commuter travel is evident each weekday in metropolitan areas across the United States. Traffic reports dominate the radio during the morning and evening commuter peak. In public polls traffic congestion is almost always labelled one of the top problems in major urban areas. Congestion is clearly defined as a problem of loss of economic productivity instead of a symptom of an unsustainable transportation system. Meanwhile public transit, especially buses, are seen as a mode that poor people rely on, not as an environmentally friendly solution to developing a sustainable transportation system.

The failure of US transportation policy-makers and planners to adopt a framework of sustainability has also resulted in detrimental environmental impacts by consuming natural resources and causing air, water and soil pollution. Air pollution is especially prevalent in metropolitan areas, and land consumption for infrastructure and the loss of open space continues to be a problem resulting from an intensive 50-year period of highway construction, with the sprawling land use patterns that followed the highways. Policy-makers have tried to fix some of the environmental problems with technological solutions, such as cleaner engines and better emission controls, instead of supporting effective alternatives. This misguided policy direction has supported a transportation system that continues to ensure that the car is the primary means, and increasingly the only means, of transport for many trips.

However, community organizations with strong principles of social and environmental justice are helping decisionmakers understand that the methods to achieve a sustainable transportation system are the same methods that will lead to a more *effective* transportation system. Environmental justice is defined by the Principles of Environmental Justice adopted at the first People of Colour Summit in 1991.[1] This broad-based framework for understanding public policy in terms of community issues has been vital in moving forward and critiqueing policy from the perspective of low income communities and communities of colour. President Bill Clinton adopted an Executive Order on Environmental Justice in 1994 and the order has been an important tool for communities, especially low income communities and communities of colour, to impact public policy made by public agencies at state and federal level (for a discussion of environmental justice see Bullard; 1990, Bryant and Mohai; 1992; Low and Gleeson, 1998: 105–20).

Through local grassroots involvement, demanding a democratic process, communities are changing the way transportation projects are evaluated and ultimately how local, state and federal transportation policy is created in the United States. Communities are challenging public agencies to change their planning and policy setting agenda. Better public transit is being defined as part of building better communities, the myth of expanding highway capacity to reduce congestion is being revealed. Studies on increasing highway capacity are showing that this capacity is often again congested after a few years, doing nothing to relieve congestion (Fulton, 2001; Noland, 2000; Hansen, 1995; STPP, 2001). A focus on transportation for those most in need is being prioritized to improve the transportation system for all users. In the Bay Area, communities have pushed a comprehensive approach to transportation investment by linking transportation and land use in the Transportation for Livable Communities programme.

Specifically looking at transportation policymaking and planning, this chapter examines the impact of top-down transportation policies on sustainability in California and the San Francisco Bay Area as a case study of US transportation policy. We will consider how major transportation policy focused on highway building in the 1950s, and how policymakers responded to the resulting symptoms of pollution and energy consumption with environmental regulations laid down in the 1970s and 1980s. Then the chapter will review some recent changes in approaches to transportation policy including the Intermodal Surface Transportation Efficiency Act (ISTEA), welfare-to-work strategies, the rise of environmental justice and local county transportation funding measures. These new policies are beginning to move transportation in the United States towards a grassroots framework that improves the quality of life of communities while embracing the basic goal of sustainability in transportation.

How the commuter became king of the road

The major defining point in US transportation policy was the beginning of the interstate highway system in 1956. Pushed by President Dwight D. Eisenhower as the road infrastructure necessary then to defend the United States in case of a foreign attack, the Interstate Highway Act prioritized the funding to build a freeway system that linked cities to suburban areas. The Act increased annual spending on highways by six times and set taxes on cars and gasoline that funded the construction of the highway system (Gordon, 1991). Highways opened up new areas to build housing resulting in poorly planned suburbs that did not support public transit and non-motorized transport. Supported by unfriendly federal housing policy that encouraged developers to build housing for middle-class whites in new suburban areas, highways initially gave those who moved to the suburbs quick access to urban centers where the majority of jobs were located (Jackson, 1985). Over time these highways became congestion nightmares on weekdays, filled with commuters trying to get to and from work.

Highways resulting from this federal policy were not welcomed in many urban areas across the US. These freeways destroyed housing predominantly in communities of people of colour, and created health and air pollution problems in these communities. In the San Francisco Bay Area and in the rest of the country, communities in urban areas fought the massive highway system being built in their neighborhoods and city. San Francisco became famous for freeway revolts as communities rallied to prevent freeways from going through the heart of the city, including its biggest open space, Golden Gate Park. They succeeded in their limited aim, but still freeways cut through the edge of the city and isolated the predominantly African-American neighbourhood, Bayview Hunters Point. Just across the Bay in Oakland the fabric of the West Oakland community, another African-American neighbourhood, was devastated when the Cypress freeway divided the neighbourhood. This example of poor planning of the freeway system without any regard to the neighbourhoods they cut through illustrates the ugly social impact of highway building.

While Federal transportation policy supported building highways to urban job centres for suburban commuters, government developed new federal policies to address growing environmental problems that resulted from the increasing dependence on the car. As the low density of housing in the suburbs forced the dependency on car use and resulted in decreased transit use, the air pollution in metropolitan areas increased and smog became a major problem. Federal policy encouraged technology changes to solve environmental problems. The Motor Vehicle Air Pollution and Control Act of 1965 set strict emission standards for cars. The 1970 Clean Air Act put even stricter emission standards on motor vehicles. Further amendments to the Clean Air Act in 1977 and 1990 continued to improve emission standards, support minor behaviour changes in commuters, and promote the development of alternative fuel vehicles.

In addition, in the 1960s and 1970s federal policymakers tried to revive public transit as a viable commute option and as a means to reduce pollution. However, because of the tremendous disjunction between where people worked and where they lived, the new transit service did little to ease cities' dependence on the automobile, especially in the suburbs. The 1964 Urban Mass Transportation Act and the 1970 Urban Mass Transportation Assistance Act attempted to get people out of their cars by building commuter railways and giving federal funding support for public transit to operate. Still, urban air pollution worsened across the country due to soaring automobile use.

Then, in the 1970s, an energy crisis spurred by gasoline shortages, caused further federal policy reaction, creating policies to accelerate technological changes to cars that reduced emissions and improved energy efficiency. Corporate Average Fuel Economy (CAFE) standards were developed to improve the fuel efficiency of cars and to attempt to reduce the dependence on oil. Adopted in 1975 when the average fuel consumption for cars was less than 14 miles per gallon (5.95 kilometres per litre – kpl), this policy improved the average fuel consumption of cars to 27.5 mpg (11.7 kpl) by 1986. However, this standard has not changed since 1986, mainly due to the resistance of the automobile manufacturers and the highway lobby to make further improvements.

Unsustainable transportation trends

The focus of transportation policy on funding solutions for the peak commuter and on technological solutions to mitigate resulting environmental problems were small steps towards improving the sustainability of the transportation system in the United States. California is often looked to for innovative policy solutions and the San Francisco Bay Area is one of the more accessible transportation systems of US metropolitan regions. Nevertheless, Bay Area policy continued to focus on the peak commuter, and the worsening trends there mirrors the experience of many other regions.

First, transit use as a percentage of all weekday trips in the region has declined since 1970 and projections to 2020 continue to see its modal share

Table 6.1 Selected transportation characteristics of the San Francisco Bay Area: 1970–2020

Characteristic	**1970**	**1980**	**1990**	**2000**	**2010**	**2020**
% transit trips as a share of all weekday trips	7.1%*	6.6%	6.8%	6.7%	6.6%	6.4%
% auto person trips as a share of all weekday trips	92.9%*	79.6%	82.8%	83.0%	83.6%	84.2%
Total population: (million people)	4.631	5.180	6.024	6.824	7.397	7.774
Auto person trips (weekday: million trips)	10.287	13.566	14.964	17.017	19.366	20.799
Transit trips/ (weekday: million trips)	0.791	1.132	1.221	1.383	1.526	1.582
Vehicle miles travelled (weekday: million miles)	53.622	86.989	107.708	127.770	149.889	166.791
Vehicle kilometres travelled (weekday: million kilometres)	86.278	139.933	173.302	205.582	241.171	268.367
Total vehicles per household	1.339	1.683	1.761	1.845	1.918	1.928

Note
*The 1970 figures exclude bicycle and walk trips in the total of all trips.

Source: Metropolitan Transportation Commission.

decline (See Table 6.1). Despite a system that has broad coverage, transit trips currently only represent 6.7 per cent of all weekday trips in the Bay Area.

Auto use will continue to dominate Bay Area travel habits, with more cars, more driving and longer commuter journeys. While this trend is not as frightening as in the 1980s when the growth rate of vehicle miles travelled (VMT) was five times greater than the population growth rate in the Bay Area, projections from 2000 to 2020 show that VMT growth rate (30 per cent) will be twice that of population growth rate (14 per cent) (see Table 6.2). Households will own even more vehicles, continuing the rampant increase from the 1950s and 1960s when average vehicles per household increased from 1.2 in 1950 to 1.61 in 1970. In 2020, projections show the average number of vehicles per household almost reaching two. Even with continuing efforts to relieve highway congestion, it will increase overall in the region by over 150 per cent.

Environmental problems will not improve either. The Bay Area will continue to fail to attain Federal and State air quality standards for carbon

Table 6.2 Per cent change in population and distance travelled (VMT, VKT) in the San Francisco Bay Area, 1970–2020

	1970–1980 (%)	1980–1990 (%)	1990–2000 (%)	2000–2010 (%)	2010–2020 (%)
Population Growth	12	16	13	8	5
Growth of vehicle miles/kms travelled	62	24	19	17	11

Source: Metropolitan Transportation Commission.

monoxide where, each summer, air quality reaches unhealthy standards. While technology has improved emissions controls for cars, the increase in VMT continues to threaten the health of Bay Area communities. Moreover, continuing road building to support suburban development is leading to the loss of more open space and farmland. The need for alternatives to the car will still be widespread. Even though households without vehicles are projected to decrease from 8.9 per cent to 7.6 per cent of all households between 2000 and 2020, still over 200,000 households – along with those too young or too old to drive – will continue to be dependent on transit and other alternatives.

Signs of sustainable transportation policy

The need to develop a policy framework to address the sustainability of the transportation system is transforming policy at the local, state, and national levels. Nationally, this transformation began in 1991 with the Intermodal Surface Transportation Efficiency Act (ISTEA). Influenced by environmental groups lobbying to give local transportation agencies more flexibility to spend federal funding, ISTEA began a dialogue around changing policies to support transportation from a community perspective. This continuing change was reflected in the reauthorization debate again as the policy was transformed into the Transportation Equity Act of the twenty first century (TEA-21) in 1997.

Increasingly, policy goals are also trying to address social needs. Until recently, local, state and federal policymaking ignored the importance of public transit for the needs of low income people to access and to keep their jobs. Given the cost of owning and operating a car, policymakers are realizing that meeting the needs of the poorest must be a priority. Social service agencies are becoming stakeholders in ensuring that the poor have access to transportation through better policy that supports funding for transit and that is affordable. Thus transportation agencies are being forced to address the most vulnerable of our communities' transportation needs.

Box 6.1 ISTEA and TEA21

The major federal initiative of the 1990s, the Intermodal Surface Transportation Efficiency Act (ISTEA), was designed to integrate land use and transportation planning with an emphasis on environmental improvement and 'community' planning – as well as funding new roads (USDoT 2000: 1–2). But Title 1 of ISTEA targeted seven times more funds to highway-related projects (US$63.9 billion) than to non-highway projects (US$9.2 billion) (ISTEA: 2000: 3). ISTEA was replaced by TEA21. The Transportation Equity Act for the Twenty First Century (TEA21), which became law in June 1998, continues the ISTEA direction but with a reduced funding share for new highways. TEA21 increases funding over ISTEA by 40% over six years to US$217 billion with US$172 billion in 'guaranteed funding' for highways and US$36 billion for transit.

TEA21 provides for a 54 % decrease in the share of funding going to the construction of new highways, from 8.2% of total funds to 3.7%. Funds are provided for traffic calming, demand management, the design of new 'transit-oriented' land development projects, 'intermodal connectivity', environmental improvement and the reduction of CO_2 emissions, research into high technology transport systems such as magnetic levitation and 'intelligent systems', and new tools for using transportation 'to revitalize communities and create alternatives to driving'. One of these new tools is a fund to provide new transit options to enable inner city job seekers to get to work in the suburbs. In this respect environmental justice is introduced as a criterion for TEA21 funding – a direct appeal to social protection. New rail projects are springing up, as the User Guide observes, 'in some unlikely places' (*ibid.*). Funding is authorised for new rail projects in 26 cities including Atlanta (Georgia), Dallas (Texas), Memphis (Tennessee) and San Jose (California). Even so US$8 billion will be spent on new roads, and much more on highway related projects (see Table 6.3).

Table 6.3 Funds earmarked under the Federal Transportation Equity Act (TEA21)

Category	**ISTEA US$ billion**	**TEA21 US$ billion**
Money that must be used to build new highways (interstate construction, etc.)	13	8
Money that may not be used to build new highways (bridges and highway maintenance, transit, etc.)	74	111
Flexible money (surface transportation program, etc.)	69	97
Total	**156**	**216**

Source: United States Department of Transportation (2000) *The Transportation Equity Act, for the Twenty First Century: Users' Guide On Line.* http://www.istea.org/guide/guideonline.htm

San Francisco Bay Area: leading the way to healthy and sustainable communities

The shifting of transportation policy from meeting the needs of middle income suburban commuters to meeting the transportation needs of communities is necessary to achieve a sustainable transportation system. The Bay Area and California are seen as leaders in developing innovative transportation policy. The Bay Area is also seen as a region where many community-based advocacy groups have claimed a stake in defining how policy is set. In addition, there is a regional transportation agency, the Metropolitan Transportation Commission (MTC) that passes on funding to local transportation agencies. MTC is directly responsible for developing a Regional Transportation Plan (RTP) every 3 years which becomes the blueprint for spending in the region over 25 years.

In 1998 after the momentum gained from improved federal transportation policy and with a booming economy increasing Bay Area transportation problems, the climate was right to begin to build transportation policy from a community grassroots perspective. The RTP worth US$81 billion was undergoing a major update and the initial documents from the process showed disturbing trends: A 249 per cent increase in congestion, a decline in transit as a share of total trips and a deficit for transit maintenance of US$375 million.

A year earlier a new regional coalition of environmental groups, transportation activists and community groups had come together to begin to develop an agenda for transportation based on sustainability. The Bay Area Transportation and Land Use Coalition (BATLUC) put these transportation trends in the public spotlight through the media, calling for, among other things, full funding for public transit in the Bay Area. In addition BATLUC members began to attend public meetings in mass, write letters to raise the voice of communities, and to demand transportation options for all segments of the Bay area population.

The key to BATLUC's success was their ability to build a multi-ethnic and low income coalition of allies that could speak from a grassroots perspective and for a variety of communities. In the summer of 1998, the MTC Board was poised to pass the RTP with no changes resulting from the public comment process. With significant environmentalist support for BATLUC's 100 per cent funding for transit, BATLUC's efforts received a major boost when groups representing low-income communities starting bringing up to 100 people to key meetings, even at one location in the suburbs that had almost no transit access. The breadth and diversity of groups calling for this recommendation led the appointed commissioners to reject their own staff's proposal, and for the first time to accept the proposal from the community groups. The end result was that US$375 million was shifted to fund public transit. There was an acknowledged power shift as well: MTC's executive director one month later told his commissioners that there were new 'players'

Box 6.2 The Proposals of the Bay Area Transportation and Land Use Coalition (BATLUC) 1998 and 2000

The specific changes requested by BATLUC in 1998:

1. Adopt measurable performance goals with quantifiable targets in the Plan. The BATLUC platform called for three target indicators:
   - People should not have to drive more than they do today. In technical terms average vehicle miles traveled (VMT per capita) should not increase, and preferably should decrease over the next twenty years.
   - Transit should be convenient and effective, so that there is no decline, and preferably an increase, in transit's share of trips.
   - Walking and cycling, the cheapest, least polluting forms of transportation should be made safe and convenient so that these modes show no decrease, and preferably an increase, in trip share.
2. Develop measures of social equity with an explicit equity analysis of the Regional Transportation Plan that analyzes the effect of the Plan on low income communities and communities of color.
3. Fully fund the transit maintenance needs of transit operators so they can provide a high level of service and not need to raise fares or make service cuts. The shortfall makes up $375 million in the regional transportation Plan (AC Transit: $51.5 million, Golden Gate Transit: $19.6 million, Caltrain: $104 million, Muni: $0.8 million, BART: $99.5 million)

BATLUC Proposals 2000:
The shift of $186 million in 2000 went to:

- Expanding the night and weekend transit service ($95 million)
- More paratransit for seniors and the disabled ($44 million)
- Additional bicycle and pedestrian safety and access ($44 million)
- Beginning a new program to support transit-oriented development ($3 million).

at the table who were developing their own recommendations, and that MTC staff would in the future try to put forward policy *options* instead of a single recommendation.

A year later another major fight was brewing. In response to the local need for funding to make transportation improvements, many counties in California have passed sales tax measures for transportation funding. In Alameda County in 1998, Measure B was placed on the ballot with a list of projects that were to be funded. This list of projects split the environmental community that opposed its highway expansions and the social justice community that mostly supported the increase in transit it would support. Not surprisingly, the measure failed to gain the two thirds approval of the voters needed to pass. In 2000, a local chapter of BATLUC united the environmental and social justice groups around a common agenda, and, by negotiating as a team, was able to shift an additional $186 million of the $1.4 billion measure to transit and bicycle/pedestrian facilities. With unanimous support from all of the groups, and a grassroots campaign led by

BATLUC, the measure passed with an overwhelming 81 per cent of the electorate voting for it.

Youth are also changing the debate around transportation policy away from the needs of the peak suburban commuter. Many young people across the Bay Area do not have guaranteed transportation for the school journey. Many rely on public transit. Youth from low income families cannot always afford transit passes to take the bus to get to and from school. This inability to get to and from school has impacted many students' ability to get their education. As a result, young people are leading the way in demanding that MTC provide transit passes for youth to and from school. After many community groups in several counties pressured MTC, they agreed to provide matching funds to provide free transit passes to low-income youth.

Environmental justice issues are also being raised to increase support for public transit in low income communities. In the San Francisco Bay Area, the US Department of Transportation required MTC to better analyse the transportation needs of low income people as part of the Regional Transportation Plan process in 2001. In addition, in 2001 a lawsuit was settled that contended that MTC failed to increase public transit ridership to meet Clean Air Act requirements.

The plaintiffs framed this issue as one of 'environmental justice'. In the Bay Area, as in many other urban areas, transit operators struggle to find funding to keep services in operation. When transit operating costs are not met, typically fares are raised or weekend and night services are cut. These decisions hurt the poor and people of colour most because they use public transport most, or depend on the night and weekend service to get jobs during non-commuting hours. In this instance the failure of MTC to increase transit use by increasing the amount of funding going to transit operations hurt low income communities and communities of colour most. In the settlement, increased transit service is being targetted to these neighbourhoods.

The opportunities ahead: communities must steer policy

The examples in the last section illustrate how community groups are beginning to set the framework of policy for the Bay Area. The terms of the policy debate have changed from meeting the needs of the peak suburban commuter to meeting the needs of all segments of the population, especially those most vulnerable. These decisionmaking and policy reforms represent the kind of changes necessary to develop a sustainable transportation system.

From this overview of the San Francisco Bay Area, four key lessons can be learned that are necessary to developing a transportation policy and planning framework that will guide the United States towards a sustainable transportation system.

1. Environmental justice must be incorporated into the transportation policymaking framework. When the environmental justice movement began in 1991, the needs of people of colour and low income people were not a national priority in environmental and transportation issues. Addressing the needs of people of colour and the poor will be vital in addressing transportation sustainability. Community groups from low income communities and communities of colour have been vital to changing the transportation debate in the Bay Area. They must continue to be included in the policy development and decisionmaking process and public agencies must analyse environmental justice issues in their policymaking processes.
2. Prioritizing the transportation needs of those most vulnerable will improve the transportation system for all users. For too long the transportation needs of those most vulnerable, the poor, disabled, young and elderly have been left out of the decisionmaking process. For example, in 1996 when changes to the welfare system occurred, transportation and social service agencies discovered that poor transit options created a major barrier to those going from receiving public assistance to employment. Not only did many recipients not own a car but also public transit was not available to new job locations, or else it did not run at night or during weekends. However, in the last five years social service providers have begun to work with transportation agencies to make sure the transportation needs of this segment of the population are met. These issues have helped shift the focus of transportation funding to greater needs than just those of the peak suburban commuter. Public agencies and decisionmakers must ensure that transportation options are available to these segments of the population.
3. Public agencies must make their decisionmaking clear to the public by making data available and presenting it in an objective manner. In 1998 BATLUC was able to capture much of the attention of the media and public around the RTP because it took MTC's numbers from the RTP and translated the information in a useful way for the public. For example tradeoffs between highway projects and transit projects must be made clear by looking at the outcomes of projects. One of the major policy debates in the San Francisco Bay Area has been around extending our regional rail system to San Jose and Silicon Valley. BATLUC has fought to ensure that building this extension does not reduce funding for the bus service in the area. Transportation agencies have a clear responsibility to present the impact of such decisions.
4. Finally, public agencies must work together with communities to ensure that transportation and land use issues are connected in policy decisions. MTC has adopted the Transportation for Livable Communities program to make this connection. Hailed as a model policy for funding nationally, this programme supports projects, which connect good public transit

with land uses that are community friendly. This programme has been wildly popular and successful and is vital to developing a sustainable transportation system.

Note

1. 'People of colour' is a term typically used in the USA to describe people of non-European origin whose ethnicities include (but are not limited to) African-American, Asian-American, Latino, Middle Eastern and Native American. The term is commonly used because of the racism that many people of non-European heritage have faced historically and continue to face today in the USA in both individual and institutional settings.

7
Transport in the European Union: Time to Decide

John Whitelegg

Introduction

In October 2001 the European Commission, the administrative arm of the European Union (EU), published its new transport policy document with the sub-title 'Time to Decide' (EC, 2001 http://europa.eu.int/comm/energy_transport/en/lb_en.html). Clearly the Commission has already decided that transport conditions in Europe for over 300 million citizens are rather poor now and are set to deteriorate even further by the year 2010, which is the time horizon of the transport policy document. Sadly the Commission set the whole transport policy debate within a very unimaginative context of some modest modal shift (less use of road and more use of rail) and some suggestions for charging lorries (road haulage vehicles) for their use of road space.

This misses the opportunities for fundamental demand management strategies (see Chapter 14). The Commission finds it very difficult to grasp the concept of traffic reduction. We can reduce the numbers of cars on the streets by land use planning and demand management. We can reduce the number of lorries (and the distances they travel) by innovative spatial/regional organizational strategies designed to increase the proportion of local products consumed locally and we can reduce the explosive increase in demand for air travel by making sure that the cost to the passenger 'tells the ecological truth' and that airports respect elementary public health controls. All this has been missed by the Commission and its White Paper now sets us very firmly on a 'business as usual' trajectory with just a little bit of hand waving and concession to the 'green lobby'. The European Commission has decided to do a lot more of the same thing over the next 10 years:

- The consequences of pursuing a 'business as usual' strategy are very clear.
- Increasing levels of congestion on Europe's motorways and urban roads. Congestion costs alone in the EU are expected to equal 80 billion Euros per annum.

- Significant increases in greenhouse gases (GHG) from transport. Transport's production of GHG will make the achievement of EU reduction targets very difficult to achieve.
- Significant health impacts from lack of physical exercise, noise, respiratory disease and road traffic accidents.
- Significant damage to neighbourhoods, community and liveability as traffic levels damage social interaction and public use of streets.
- Significant costs in futile attempts to cope with the steep increase in transport demand. The 10 year transport plan of the United Kingdom (UK) calls for £180 billion investment in physical infrastructure.
- More polarized societies as wealthier groups travel more (air, high speed train and car), poorer groups travel less and the environmental impacts of travel on the part of the wealthy bear down disproportionately on the poorer groups.
- Spatial changes in European society so that urban sprawl accelerates, cities lose their cultural and economic significance and move in the direction of 'doughnut cities'.
- Agricultural land, land for nature protection and recreational land is lost and travel distances increase.

The growth in demand for transport

The White Paper has described the main elements of the transport problem. Most passenger and goods traffic goes by road. In 1998 road transport accounted for nearly half of all goods traffic (44 per cent) and 79 per cent of passenger traffic. The road share of freight transport is much higher than the 44 per cent figure if we look at land transport by itself. In the UK, for example, road freight carries 70 per cent of all freight tonne kilometres. The figure reduces on a European scale because of the importance of coastal shipping and freight on the inland waterways (especially the Rhine and the Danube).

In the period 1970–2000 the number of cars in the EU trebled from 62.5 million to nearly 175 million. Overall, car travel (car-kilometres) is growing at about three per cent per annum. The number of cars is rising by 3 million every year and the enlargement plans of the EU which over the next 10 years will bring Poland, Hungary and Romania (and probably others) into the Union will add to this total. The so-called accession countries in common with many other ex-Soviet bloc countries (the Baltic states, Slovakia, the Czech Republic) are all experiencing car ownership and use growth rates that are much larger than the current EU average. These growth rates are encouraged by the EU through its funding mechanisms for encouraging motorway construction (e.g. the motorway around Budapest) and pursuing grandiose plans for long distance motorway connections (Trans European Networks: see below).

Road haulage is set to grow by about 50 per cent by 2010. In the period 1970–2000 it grew at about 5 per cent per annum. Changes in European manufacturing organization, logistics and sourcing strategies are all adding to the burden of road freight. Road freight grows as a result of the organizational decisions of production companies and logistics operations and not because of the growth in the demand for physical products. The distances over which goods move every year grows by around 6 per cent and the decisions of manufacturers to move those goods over thousands of kilometres (because transport is too cheap) fuels the demand for new motorways, new Alpine crossings and new crossings between France and Spain (the Pyrenees). The EU is determined to assist in this process of energy-intensive, distance-intensive production development which is why there is so much emphasis in the White Paper on very expensive new infrastructure to connect Germany and Denmark with a fixed link (the Fehmarn crossing). The cost of the EU's planned major infrastructure links is a staggering 1800 billion Euros and much of this capacity will simply move products around from region to region when they are available in all regions anyway. This has been dubbed 'pointless transport' (Whitelegg, 1994).

There is even faster growth in demand for aviation than for car travel or road freight. Passenger kilometres flown are growing in Europe by about 8 per cent per annum and the very exceptional and tragic circumstances of 11 September are very unlikely to damage the medium to long term growth potential of the aviation industry. Indeed in late November 2001 the UK government gave approval for the construction of Heathrow Terminal 5, which at 30 million passengers per annum capacity is bigger than Frankfurt Airport and will mean that Heathrow can handle 100 million passengers per annum by 2010–15. This decision comes at the request of the aviation industry and is matched by similar decisions to build new airport capacity, including yet another completely new airport, for Paris. The aviation industry is clearly very confident about future success even though it is currently taking advantage of the unique circumstances surrounding international terrorism and is making staff redundant and reducing flights. This will merely increase the amount of public subsidy (£40 million was given to the aviation industry by the UK government in December 2001) and restore the usual trajectory which in the UK will require a new airport equal in size to Heathrow somewhere in the country every 10 years.

The European Commission, having expressed concern about greenhouse gases, congestion, land take and other consequences of the growth in demand for transport then goes on to say that the growth of aviation is 'inevitable':

> This reorganisation of Europe's sky must be accompanied by a policy to ensure that the *inevitable* expansion of airport capacity linked, in particular, with enlargement remains strictly subject to new regulations to reduce noise and pollution from aircraft (EC, 2001: 14; emphasis added).

This is the clearest expression of the Commission's fundamental lack of understanding of sustainable development and transport. There is nothing inevitable about the growth in demand for transport, including growth in demand for aviation.

Transport demand and spatial re-organization

The main source of growth in demand for transport in Europe is the spatial reorganization of society to produce a more spread out 'distance-intensive' life style, a concentration of facilities (e.g. retailing) in fewer locations and an increase in the space-time complexity of everyday activities (e.g. two working adults and children going to different schools beyond walking and cycling distance). Transport planning and infrastructure provision cannot compensate for fundamental inefficiencies in spatial organization and service delivery. The practical experience of South-East England is that the road network cannot cope with the demands made on it as commuter trips increase in length and school children are taken to school by car. The most serious congestion 'hot spots' in Europe (around Paris, Frankfurt and London) are the result of spatial changes in the organization of services and facilities, and solutions are more likely to be found in re-engineering those spatial changes and not in providing more roads.

John Roberts's comparison of Almere (The Netherlands) and Milton Keynes (England) demonstrated the extent to which land use and transport planning can influence the demand for motorized transport: 'the most obvious finding and an important one, was the much higher percentage of trips made by car and the much lower level of bicycle use in Milton Keynes when compared to Almere (65.7 per cent of trips by car compared to 43.1 per cent, 5.8 per cent of trips by bicycle compared to 27.5 per cent respectively)' (Roberts, 1991: 1). The influence of compact cities on reducing motorized trips is reviewed in Smith *et al.* (1998). Physical land use planning is a tried and tested method of reducing the length of trips, increasing the use of non-motorized modes and reducing the demand for expensive road infrastructure.

Greenhouse gases

Emissions of CO_2 from transport in the EU increased by 47 per cent between 1985 and 2001. Other sectors increased by 4.2 per cent. More than 30 per cent of final energy in the EU is now consumed by transport. If this trend continues the EU will not meet its Kyoto commitments. Road transport is the main cause of this increase and contributed 84 per cent of the CO_2 emissions from transport in 1998. Emissions of CO_2 from road freight are also expected to rise substantially, by 33 per cent between 1990 and 2010. Road transport is also a small but growing source of nitrous oxide (N_2O) emissions from passenger car catalysts. Emissions doubled between 1990 and 1998 to 7 per cent of total N_2O emissions (TERM, 2001).

In 1998 EU greenhouse gas emissions from international transport (aviation and shipping) amounted to 5 per cent of total EU emissions. Aviation emissions are expected to rise dramatically in future years (Whitelegg and Williams, 2000) and to account for about 15 per cent of greenhouse gas emissions from the transport sector by the year 2020.

Attempts to restrict or reduce GHG from transport in Europe have largely failed. A voluntary agreement with the automobile industry to set an emission limit of 140 g/km of CO_2 will reduce the rate of growth by a few percentage points but is more than compensated for by the increase in popularity of sports utility vehicles (SUVs). SUVs are very large, heavy, four wheel drive, jeep like vehicles with very poor fuel efficiency and very high GHG emissions (>250 g/km). Their market penetration is very high in the US and in Europe. Average vehicle occupancy is falling as more people choose to travel alone in their vehicles thus increasing per capita GHG and cancelling out fuel efficiency gains. Put very crudely European car transport is characterized by increasingly fewer persons in increasingly heavier and more polluting vehicles. This is not a trend that can be influenced by technology.

Urban sprawl and logistic tendencies also exert a powerful influence on GHG emissions. The UK has seen a substantial increase in car use for retailing as retailing has moved towards the US style 'shopping mall' concept. Large shopping centres on edge of city sites (e.g. Meadowhall in Sheffield, Metro Centre in Newcastle and Trafford Centre in Manchester) have all proceeded on the basis of attracting shoppers from distances of over 100 kms to their thousands of car-parking places. The impact on traditional retailing centres has been very negative, for instance on Dudley near Meadowhall which has lost about 40 per cent of its retail space, and shopping trips account for more miles travelled than commuting trips.

Other European countries have resisted these trends. In Denmark shopping is still conducted in traditional city centre retail outlets supported by local residents. Germany has also resisted these trends but the accession states (Hungary, Poland, etc.) are moving very fast in the direction of the out of town centre. Urban sprawl has the potential to counteract and overcompensate for any technological gains made in fuel or engine efficiency and also any gains made by fiscal means (e.g. fuel taxation, congestion charging, car park charges). Without a radical change in the business/developer driven agenda towards business parks, shopping malls and new models of suburbia, Europe will not meet Kyoto targets and will see a continuing growth in transport GHG emissions.

The European Environment Agency in Copenhagen has addressed this issue and predicted that GHG emissions from transport will be 39 per cent above the 1990 levels by 2010 (EEA, 2001). This does not include the aviation GHG emissions, which can be allocated to EU citizens on the basis of their air miles and are currently not counted in EU GHG inventories.

Road freight transport is a particular source of concern in the EU. Road freight tonne/km increased by 29 per cent in the period 1990–97 and road freight could increase by much more in the future. The road freight problem is essentially one of the spatial organization of production and logistical supply chains. Paradoxically, because logistics is so well organized and so sophisticated (just in time, very short lead times, satellite tracking, reliability), it is now possible in Europe at very low costs to source a huge variety of raw material and semi-processed inputs into a production chain that is very widely dispersed. Essentially the traditional barriers of the friction of distance and the cost of movement have been removed. It is now normal to move thousands of products over tens of thousands of kilometres in a highly efficient manner. The products will be delivered exactly where they are needed at the time they are needed. One of the best documented examples of this process is the case of the yoghurt pot (Böge, 1995). Böge made a study of yoghurt production at one factory in Stuttgart (Germany) and found that many different products and sub-products went into making the final consumer product. The final product was so transport intensive that it was possible to calculate that each 150 g pot of yoghurt was responsible for moving one lorry 9.2 metres. Similar trends can be observed throughout food retailing (onions from Poland) and globally (onions on sale in UK supermarkets from New Zealand). Furniture and household products sold by IKEA throughout the EU are made up of parts sourced in eastern Europe (e.g. Poland), assembled at various place in Europe, shipped to Sweden, shipped to the UK and so on. The distance intensity of such production processes is increasing at an increasing rate and produces ever more GHG.

The health impact of transport

It is only very recently that the full extent of transport's negative impact on health has become clearer. In an ecological audit of the impact of cars on German society Teufel *et al.* (1999) concluded that cars were responsible for 47,000 deaths each year and a range of other, less severe, health impacts. These are summarized in Table 7.1.

The volume of death and illness revealed in Table 7.1 puts the European transport problem into a very serious public health perspective. Transport is a major health problem and should be tackled as much within a public health context as in a traditional transport/roads/highway context. All the deaths and injuries in Table 7.1 relate only to cars. Total deaths are about 5 times greater than deaths from road traffic accidents. The health impact of various types of air pollution from transport is shown in Box 7.1.

The total amount of sickness, days in hospital etc. imposes a huge burden on the health services of European countries and this burden is not recovered from those who drive cars. The health impact is a vast human tragedy. A total of 15 million days of use of bronchodilators represents a huge problem for

Table 7.1 Health damage caused by cars, Germany, 1996 (annual totals)

Mortality	**No.**	**Unit**
Deaths from particulate pollution	25,500	deaths, per annum
Deaths from lung cancer	8,700	deaths, per annum
Deaths, from heart attacks	2,000	deaths, per annum
Deaths from summer smog	1,900	deaths, per annum
Deaths from road traffic accidents (RTAs)	8,758	deaths, per annum
Total	47,000	deaths, per annum
Injury and disease		
Serious injuries (RTAs)	116,456	injured per annum
Light injuries (RTAs)	376,702	injured per annum
Chronic bronchitis (adults)	218,000	number of illnesses per annum
Invalidity due to chronic bronchitis	110	number of invalidates per annum
Coughs/phlegm (Auswurf)	92,400,000	days per annum
Bronchitis (children)	313,000	number of illnesses per annum
Repeated coughing (Wiederholthusten)	1,440,000	number of illnesses per annum
Hospitalization with breathing problems	600	number of hospitalizations per annum
Hospitalization with breathing problems	9,200	number of days of care per annum
Hospitalization (cardiovascular disease)	600	hospitalizations per annum
Hospitalization (cardiovascular disease)	8,200	number of days of care per annum
Incapacity for work (Arbeitsunfähigkeit) not including cancer	24,600,000	days per annum
Asthma attacks (days with attacks)	14,000,000	days per annum
Asthma attacks (days with broncho-dilator)	15,000,000	days per annum

Source: Teufel *et al.* (1999), *passim.*

many children and many families and the impact on physical activity, social activity, enjoyment of outdoor pursuits, community and neighbourhood is incalculable. Health impacts in Europe in the twenty-first century are the direct equivalent of disease impacts in nineteenth-century cities (cholera, typhoid), which then required major re-engineering with clean drinking water and sewage systems. We are still waiting for the twenty-first century equivalent of this re-engineering to deal with the modern equivalent of widely dispersed sewage.

Box 7.1 Impact of human health of air pollutants from transport

Carbon Monoxide (CO): at high levels causes headaches, drowsiness, nausea, slowed reflexes and at very high levels, death. At low levels it can impair concentration and nervous system function and may cause exercise-related heart pain in people with coronary heart disease

Nitrogen Oxides (NO_x): impairs respiratory cell function and damages blood capillaries and cells of the immune system. This effect in turn may increase susceptibility to infection and aggravate asthma, In children exposure may result in coughs, colds, phlegm, shortness of breath, chronic wheezing and respiratory diseases including bronchitis.

Ozone (O_3): reduces lung function in healthy people as well as those with asthma. It may increase susceptibility to infection and responsiveness to allergens such as pollens and house dust mites. It may cause coughs, eye, nose and throat irritation, headaches, nausea, chest pain and loss of lung efficiency. It increases the likelihood of asthma attacks.

Particulate matter (PM): strongly associated with a wide range of symptoms such as coughs, colds, phlegm, sinusitis, shortness of breath, chronic wheezing, chest pain, asthma, bronchitis, emphysema and loss of lung efficiency. As many as 15 per cent of asthma and 7 per cent of Chronic Obstructive Pulmonary Disease cases in the urban population are estimated to be possibly related to prolonged exposure to high concentrations of PM. Long term exposure is associated with increased risk of death from heart and lung diseases. PM may carry carcinogens such as polycyclic aromatic hydrocarbons (PAHs), hence may increase the risk of developing cancer.

Volatile Organic Compounds (VOC): This category of pollutant includes thousands of different chemicals many of which are hydrocarbons (HC). They may cause skin irritation and breathing difficulties; long term exposure may impair lung function. Many individual compounds are carcinogenic (including benzene). Benzene can cause leukaemia. Those most at risk are people exposed to benzene at work or who live or work in the vicinity of petrol filling stations or general vehicle activity.

Sulphur Dioxide (SO_2): irritates the lungs and is associated with chronic bronchitis. People with asthma are particularly vulnerable and a few minute's exposure to the pollutant may trigger an attack. However the most serious effect occurs when SO_2 is absorbed by particulate matter which is then inhaled deep into the lungs. At high doses it can release sulphuric acid on reaction with moisture in the lungs. This can result in widespread death and illness, for example, it is likely to have been the main cause of the 4000 deaths during the notorious 1952 London smog.

Source: Dobson-Mouawad *et al.* (1998), *passim.*

Road traffic noise and noise from aircraft also create significant health problems (WHO, 1996). These health problems are generally understated in Europe with an implicit assumption on the part of traffic engineers and planners that most people can get used to noise and, in any case, it is only a minor irritation and part of life in an advanced industrial society. This has to be rejected. Noise causes raised blood pressure, cardiovascular disease, a range of psychological problems and sleep disturbance, and it damages school age children if they are exposed to noise in a learning environment. WHO discusses the evidence supporting the contention that children exposed to noise learn less well and have reading abilities lower than is the case for children learning in quieter environments (WHO, 1993). Studies around Heathrow Airport in South East England also point to damage to children living near the airport and under flight paths (Whitelegg and Williams, 2000).

Children suffer in other ways as a result of the growth in car use and mobility. It is common in the UK, though less so in Germany, Denmark and the Netherlands for children to be taken to school by car. Hillman *et al.* (1992) drew attention to the serious impact of this tendency especially in the loss of independent mobility on the part of young children. The consequences of this loss of independence and physical activity are that the UK has the highest rate of obesity amongst its 15 and 16 year olds in the European Union (NAO, 2001). Children (and young adults) increasingly living a sedentary lifestyle with very little physical activity incur a health penalty. As they grow into full adulthood they are more likely to experience cardiovascular problems and specific illnesses such as diabetes. A National Audit Office investigation in Britain (NAO, 2001) has identified the importance of walking and cycling as an important mechanism for reducing illness, reducing demands on the National Health Service and reducing the size of the growing bill for health care.

Traffic also damages community life and it is surprising that the frequently articulated comments of urban residents in European cities about the damage to neighbourhoods, community, social interaction and 'liveability' are so poorly researched and understood. In European transport we know far more about the skid resistance of different road surface materials than we do about how traffic deeply affects psychological and physical wellbeing in urban communities. The outstanding exception to this general rule is Donald Appleyard's work in San Francisco (Appleyard, 1981). Appleyard shows in a series of diagrams that heavily trafficked streets seriously impede social interaction to the extent that residents on these streets have much less social contact and fewer friends and acquaintances than do residents on lightly trafficked streets.

This is not just an item of passing sociological interest. Isolation is keenly felt by elderly people and by parents with young children. Poor physical conditions reduce the attractiveness of urban living and contribute to economic decline, outmigration and the downward spiral of urban decay. Low levels of physical use of public space (few people actually walking) increase the likelihood of crimes against the person and burglary. Cities are attractive when they are well used by people and cyclists (as is the case in Copenhagen) or very well 'policed' in the sense that there is a significant degree of general public surveillance as is the case with German and Austrian tram systems. Travelling by tram through Dortmund or Bochum in Germany or Vienna in

Austria provide ample opportunities for everyone to survey and 'oversee' everyone else. The significant public presence is qualitatively and quantitatively different from that provided by car users who occupy a very private and insulated world.

Studies of individual exposure to pollution show that car occupants are exposed to between two and four times more pollution from vehicles than are cyclists (Rank *et al.*, 2001). This finding is in some ways counter-intuitive and surprising but is the result of cars following a similar path through traffic to that followed by all other cars and effectively driving in a 'tunnel of pollution'. This leads to the very interesting and important conclusion that the car itself damages the health of car occupants. The conventional view is that cars are safer and more pleasant than cycling which is presumed to be a dangerous activity. Scientific research shows that this is not the case and the growth of car use in Europe (especially the increase in the number of children carried around by car) represents a significant public health problem. In this case there is a direct correspondence between perpetrator and victim. Those that cause the problem suffer the consequences of that problem.

Ground level impact of aviation

In October 2001 a group of residents living in the vicinity of London Heathrow Airport won a court action before the European Court of Human Rights in Strasbourg. The Court ruled that they had been deprived of their human rights because they could not have a good night's sleep. The Court also ruled that the UK government had denied the residents an adequate remedy for their complaint and, further, that the UK government had not demonstrated that there was an overriding economic reason why night flights had to take place. This ruling is very significant indeed. It establishes a link between transport impacts and human rights that is of wider significance than just airports. Local residents in Austria, Italy and France have equally serious problems living in the vicinity of major trans-alpine motorways. The ruling widens the transport debate far beyond the boundaries of traditional studies of noise and air pollution and it demonstrates the need to adopt fundamental traffic reduction strategies. Many transport problems cannot be solved by technology or tinkering around the edges. In a crowded continent like Europe transport problems require fundamental solutions that produce reduced levels of demand.

As well as being a significant source of greenhouse gas, aviation is a very polluting activity with specific impact on urban populations. US research shows that air pollution from cars and industry has declined with time while aircraft continue to emit more ground level ozone precursors (Volatile Organic Compounds or VOCs and nitrogen oxides or NO_X) with each passing year (Natural Resources Defense Council, 1996). Airports in the US are among the top four largest emitters of NO_X and VOCs (depending on location), together with power plants, the chemical industry and oil refineries. These data are not readily available in the UK where published information (e.g. publications of the Environment Agency) does not list airports. Airports are also significant traffic generators, freight distribution centres, taxi destinations and bus stations and are responsible for significant amounts of pollution from the exhaust emissions of land based transport. They also have large amounts of fixed and mobile generating equipment to supply aircraft with power while they are on the stand and large scale maintenance facilities for engines and aircraft. They are also large fuel depots with storage tanks, fuel lines and refuelling facilities all contributing evaporative emissions of VOCs to the atmosphere.

Heathrow Airport, like many major airports, is situated in an urban area – within Greater London. Evidence presented at the public inquiry into Heathrow Terminal 5 (T5) showed that if T5 were built NO_x levels would be 110 per cent higher in 2016 compared with 1991. Without T5, levels would be only 45 per cent higher in 2016 compared with the same base year. This is a very significant increase (from an already high base line) in a pollutant which is directly associated with smog formation and with damaging human health impacts. The other pollutants (VOCs, CO and SO_2) also show increases even without T5. The London Borough of Hounslow which borders on Heathrow Airport and is responsible for air pollution monitoring is of the view that 'further expansion of the airport and associated road traffic congestion could lead to significant worsening of local air quality' (www.hounslow.gov.uk/es/monitor.html). In a press release dated 10.8.99 the same London Borough

Table 7.2 Per cent contribution of Heathrow Airport to annual emissions of four pollutants in the near Heathrow region (1991 and projected to 2016)

Case	NO_x	CO	VOC	SO_2
1991	59	45	48	76
2016 4T 50 mppa	76	57	46	66
2016 4T 60 mppa	77	57	46	67
2016 5T 80 mppa	81	63	53	73
2016 5T 100 mppa	82	61	53	74

Notes: The near Heathrow region is an area 8 km × 6 km centred on the airport, an area close to but excluding the airport itself; 4T = four terminals (the present number); 5T = five terminals; mppa 5 million passengers per annum.

Source: Evidence presented by Tabitha Stebbings on behalf of the local authorities opposed to T5 (LAHT5) at the public inquiry into Terminal 5.

concludes 'It is clear that the use of motor vehicles and the operation of Heathrow Airport heavily influence the levels of air pollution in Hounslow.'

Environment Agency data on local pollution show that emissions from Heathrow by 2016 will make it the second largest polluter for VOCs in England and Wales, after BASF (the transnational chemical corporation) on Teesside (North East England). Heathrow contributes about 10 per cent of the England and Wales total of VOCs and yet does not figure in the Environment Agency list of point sources and is not controlled on a site basis. Airports in the UK are specifically excluded from the provisions of Integrated Pollution Control. US data show that Kennedy Airport is the largest source of NO_x in New York and the second largest source of VOCs.

Both these chemicals (NO_x and VOCs) combine to form ground level ozone which in turn damages the respiratory system of humans and causes breathing difficulties, increased mortality and increased hospital admissions. VOCs emissions include a number of toxic pollutants which in addition to their role in ozone formation at ground level have a direct impact on human health. These include formaldehyde, benzene and 1,3 butadiene. A 1993 study carried out by the US Environmental Protection Agency (EPA) concluded that these pollutants contributed to elevated incidence of cancer in the vicinity of Midway Airport (SW Chicago). Midway's arriving and

Table 7.3 Zürich Airport: emissions to atmosphere

	Airport regional perimeter[1]		**Canton of Zürich**[2]	
	No_x	**HC**	**NO_x**	**HC**
Air traffic (%)	475 t[3] (60)	380 t (50)	950 t (51)	415 t (43)
Ground transport vehicles (%)	105 t (13)	50 t (7)	705 t (38)	220 t (23)
Other activities (e.g. generators, plant and machinery) (%)	210 t (27)	330 t (43)	210 t (11)	330 t (34)
Total airport (%)	790 t (100)	760 t (100)	1865 t (100)	965 t (100)
Total emissions (all sources)	2,800 t	N/a	21,160 t	39,390 t
Contribution of airport (%)	28		9	2.5

Notes

[1] The airport regional perimeter is defined as an area 9 × 12 km around the airport.

[2] The Canton of Zürich is taken to mean the area covered by the internationally defined landing and take off or LTO cycle.

[3] t = tonne.

departing planes contribute far more of these toxic pollutants than other industrial sources within a pre-defined 16 square mile study area. The EPA study estimates that aircraft engines are responsible for 10.5 per cent of the cancer cases in SW Chicago caused by toxic air pollution. There are no studies of cancer incidence and toxic pollution around UK airports, several of which are much larger than Midway.

The official view of the UK government expressed in its evidence to the T5 inquiry is that aviation contributes very little to local air pollution. The US data quoted above clearly contradicts this view and local inventories of emissions around Zürich Airport and Stockholm's Arlanda Airport show that aviation contributes a significant share of total emissions within a well-defined geographical area. The information for Zürich and Stockholm is available because both these airports are capped in terms of the pollutants they can produce. The Zürich data is presented below in Table 7.3.

The evidence on air pollution around major European airports is very clear indeed. Airports are significant contributors to air pollution and to elevated levels of particulate pollutants that are known to cause damage to human health.

Trans European networks

Europe is unique on a global scale in that it has a supra-national authority (the European Union) planning for transport and providing resources for transport projects. The 15 countries that are currently members of the European Union are still responsible in every sense for their transport plans, policies, programmes and strategies and they also make strong inputs to the EU policy level. At the European Union level there are a number of key transport policy areas:

- Reducing greenhouse gas emissions through voluntary agreements with car manufacturers.
- Improving air quality through directives and regulations which are legally binding in all member states.
- Social regulations to protect the health of workers, e.g. limiting the hours that can be driven by lorry drivers.
- Planning and funding major items of new transport infrastructure (Trans European Networks).
- Opening up access to railway systems and making national rail systems capable of being 'inter-operable'.
- Opening up road haulage to international competition through 'cabotage', i.e. permitting hauliers from one country to pick up loads in another country.
- Fiscal measures to bring the costs of road freight transport into line with the damage that is caused by lorries (i.e. increasing taxation through infrastructure charging).

In this section I will discuss the Trans European Networks (TENs) only. Trans European Networks are significant for a number of reasons. They are a key area of European Union policy in that they are intended to deliver broader political objectives. TENs are expected to contribute to the process of enlargement through easing movement across borders. TENs are expected to deliver a fully integrated political union through the binding together of the many remote and disparate regions of the Union and TENs are expected to assist in the stimulation of economic growth which, in its turn, is expected to remove economic and social inequalities in the Union. The fact that all these wider objectives and motivations lack any sound basis of evidence or validation has never been an obstacle to their promotion as key objectives.

The European Union is essentially an economic power bloc locked into outmoded ideas of economic growth, growth in output and growth in mobility. The tension at the heart of the Union is a paradoxical one of commitment to ever higher rates of growth in mobility sustained by infrastructure investment and a parallel commitment to environmentally sustainable development (see Chapter 13). By definition, sustainable development is about the conservation of energy and resources and is incompatible with commitments to the growth of motorised transport. More than any other policy area of the European Union, the commitment to TENs reveals a fundamentally unsustainable core set of beliefs and policies. The European Union is structurally incapable of adopting an economic, social, spatial or transport policy that can deliver reductions in traffic growth. The consequences of this historic missed opportunity are increasing levels of congestion, health damage, climate change and damage to the cultural and architectural fabric of Europeans towns and cities.

The European Union White Paper on transport is very clear indeed on TENs: 'Given the saturation of certain major arteries and the consequent pollution, it is essential for the European Union to complete the trans-European projects already decided' (p. 14). The projects already decided are a list of 14 separate schemes agreed at the Essen Council meeting in 1996. They are:

1. high speed rail and combined transport north–south link;
2. high speed train Paris, Brussels, Cologne, Amsterdam, London;
3. high speed train south;
4. high speed train Paris, Eastern France–Southern Germany (including Metz–Luxembourg);
5. conventional rail/combined transport Betuwe line (The Netherlands);
6. high speed rail and combined transport France–Italy (Lyon–Turin–Milan–Venice–Trieste);
7. Greek motorways PATHE and Via Egnatia;
8. multimodal link Portugal–Spain–Central Europe;
9. conventional rail link Cork–Dublin–Belfast–Larne–Stranraer;
10. Malpensa Airport (Milan, Italy);

11. fixed link Denmark–Sweden (Öresund link);
12. Nordic Triangle;
13. Ireland–UK–Benelux road link;
14. west coast main line (UK).

Three of these have been completed (9, 10 and 11 above) and the remainder will be completed by 2005. The EU budget allocated to these projects is 1830 billion Euros in the period 1995–2005. In the period 1996–97 EU funding represented 30 per cent of the total cost. The remainder of the funding was from national governments and private companies.

These huge sums of money are all devoted to encouraging European goods and people to travel further and faster. The large sums of taxpayers' money that goes into these projects has not been evaluated in terms of 'best value'. The EU is silent on the nature of the mechanisms that will convert more road space and higher rail speeds into jobs, prosperity and sustainability. Indeed the weight of scientific evidence on European transport points to the conclusion that these investments will simply stimulate more freight and passenger movement. This was predicted in a European Commission report on completing the internal market. The report *1992, The Environmental Dimension* concluded that the removal of barriers to trade and completing the internal market would increase greenhouse gases and lead to a deterioration in air quality (EC: undated).

The Gothenburg European Council (2001) concluded that TENs should form part of an environmentally sustainable policy to encourage the use of 'environment-friendly modes of transport'. This will be realised through the development of multi-modal corridors and high speed trains. Sadly the decisionmakers have chosen to ignore the reality of European transport experience. Adding to motorway capacity does not reduce vehicle kilometres travelled. Building high-speed train routes does not encourage less travel (even air travel continues to grow as air traffic control slots are released for other routes). Building airports does not reduce greenhouse gases.

TENs are increasingly concentrating on overcoming major 'natural barriers', e.g. Alpine crossings, crossing the Pyrenees and linking Denmark and Germany (Fehmarn Belt). Alpine crossings are a major political and environmental problem in Europe. The concentration of 20,000 lorries per day on key roads and tunnels has imposed intolerable burdens on people and ecology and has contributed to the breakdown of this system in a series of accidents blocking tunnels in 2000 and 2001 (e.g. the Mont Blanc tunnel). The emphasis on new rail tunnels funded by the Swiss government (not in the European Union) will only provide temporary relief to what is a fundamental spatial and organizational inefficiency.

As long as northern Europe seeks to provide southern Europe with the same kind of goods that move in the opposite direction we will always have an 'Alpine transit problem'. The problem is created by dysfunctional production systems and cost-driven marketing and consumption systems which encourage as much movement of goods as is necessary to maximise profits and reduce consumer prices. The environment (and health) is not priced in this system and all opportunities for local production for local consumption are enthusiastically and recklessly foregone. It is the express purpose of new infrastructure (road and rail) to support an illogical and unsustainable long distance production–consumption system.

The White Paper highlights the importance of new infrastructure across the Pyrenees to assist trade between the Iberian peninsula, France and countries beyond France. Currently 15,000 lorries cross this boundary each day and this number is growing at 10 per cent per annum. The European Union in advocating more crossings is not addressing the 'Why?' question. Why is lorry activity increasing so much when final consumers are not consuming 10 per cent more food, metal, plastic, wood etc. each year? The answer (once again) lies in the production system. Each year the distance intensity of each tonne of material increases. This concept is clearly described in the case of yoghurt in Böge (1995) and in Whitelegg (1997). If we continue to eliminate local production for local consumption from our economic systems then we will need to cater for very large increases in lorry numbers at every mountain pass, river crossing and international crossing. The Commission's answer is to ignore the question and to prepare plans for another high capacity rail crossing of the Pyrenees.

The White Paper suggests the addition of four new, very expensive, infrastructure projects to the TEN list:

- East European high speed train, combined transport to link Stuttgart–Munich–Salzburg–Linz–Vienna;
- Fehmarn Belt to link Germany and Denmark with a new 19 km fixed link (bridge and tunnel);
- Straubing–Vilshofen to improve navigability on the Danube;
- interoperability of the Iberian high speed rail system.

The high priority given to TENs guarantees that a high level of resources will be allocated to these projects. Europe is now very firmly established on a trajectory that will create a distance-intensive and movement-intensive society for both passengers and freight. The hyper-mobile society that we are now creating with massive subsidies and investments is already altering the geography and geometry of Europe and undermining environmental, ecological and social structures.

Conclusions

Europe is undergoing a radical transformation of its transport landscape. The process is well underway and may be unstoppable. The main elements of the change can be seen in American and Australian transport systems.

Distance is becoming a desirable commodity in itself and the fragmentation of production and consumption systems based on a false model of economic efficiency is taking us to higher levels of people and goods movement. Europe is becoming a continent where most people are on the move most of the time and if we can buy an apple that has travelled 1000 km then this must be much better than one that carries with it thousands of years of history, culture and taste and has travelled just 50 km.

The direct manifestations of transport failures (congestion, pollution and accidents) attract a response from national governments and from the European Union that will make the problem worse. Air traffic congestion is so bad in Europe that we will build 30 more airports. Road traffic congestion is so bad that we will build 10,000 km of new motorways and local travel is so awful (especially for children, those with mobility difficulties and the elderly) that we will spend 99 per cent of transport budgets on those kinds of transport that are of no relevance at all to the quality of life and accessibility of those who live in cities or the countryside.

We are creating an unequal, polarised, polluted European society where most money is spent on the irrelevant wishes and needs of the hyper-mobile and least on the quality of accessibility within 5 km of where we all live. In the process we are destroying European civilization with its fine cities and strong cultural identity and its patchwork quilt of fine landscapes. We are destroying habitat, ecology and biodiversity. We are destroying regional identity, regional food and local jobs and we are making our children ill. We are creating a polarized and divided society where the poor and the weak will be expected to absorb the environmental and health consequences of the hyper-mobility of the rich. To add a final layer of insult most of the damage is being done by governments committed to sustainable development, by businesses fully accredited to the international environmental management standard (ISO, 14001), and by a European Union that is large enough and strong enough to supply the large amounts of cash that are needed to bring about the final destruction of space through the conquest of time.

8
Transport Sustainability in Denmark, Sweden and the Netherlands

Emin Tengström

Introduction

Denmark, Sweden and the Netherlands are small European countries with well-developed environmental strategies. It is therefore interesting to consider how far they have managed to combine transport policy and environmental policy in order to achieve sustainable development. This chapter is based on my study of Danish, Swedish and Dutch transport policies in a European context (Tengström, 1999).

I will first present a geographic and demographic profile of the three countries and give a brief historical sketch of their transport policy traditions before the publication of the Brundtland Report *Our Common Future* in 1987 which really introduced the idea of sustainable development into the policy sphere (Brundtland, 1987). Following this I examine in greater detail the policies aiming at sustainable transport after 1987 and the policy successes and more pronounced failures that have occurred since then. Some explanations for these failures are canvassed and, finally, a way forward is sketched out.

The geographic, demographic and policy background of the three countries

Denmark is a small country bordered on the East by the Baltic and on the West by the North Sea, and by Germany in the South. Traffic passes through Denmark to and from Norway, Finland and Sweden. Denmark's land area consists of a number of islands and one large peninsula (Jutland). About 85 per cent of the population is urban. The population density is about 124 inhabitants per square kilometre. There is only one major metropolitan area (Copenhagen) and a few cities of medium size (Århus, Odense and Aalborg).

Sweden has a much larger land area – nearly ten times that of Denmark – and the average population density is very low (about 20 people per square kilometre). This low density results in specific transport problems for remote areas of the country. Most of the population however is concentrated in the southern region. In 1980, 83 per cent of the population was living in urban areas, many of them in middle-sized cities of between 50,000 and 120,000 population. Three of the Swedish cities, Stockholm, Göteborg (Gothenberg) and Malmö, are relatively big towns, with special problems of urban traffic.

The Netherlands is situated at the north-western edge of the European continent bordered by the North Sea in the north and West, Belgium in the South and Germany in the East. About half of the area is below sea level and the total area of the country includes some quite large expanses of water. The population density is high (472 inhabitants per square km of land area). Of the total population 90 per cent live in cities. The urban agglomeration of the Randstad (Rim City situated around a non-urban 'green heart'), including Amsterdam, the airport Schiphol, The Hague, (the national administrative capital) Utrecht and Rotterdam, with about 6 million inhabitants, is particularly traffic-intensive. The Netherlands is a transit country for large quantities of goods. The effects of the Common Market have reinforced the Netherlands' particular role of being a transit or 'gateway' country. Some basic demographic and transport data for the three countries are shown in Table 8.1 below.

According to a survey in 1993 (Danish Transport Council, 1993) Danish transport policy prior to 1987 was fairly uncoordinated. Transport policy had long been a fragmented field composed of elements such as infrastructure investment policy, land-use and other physical planning, policies concerning licence fees and duties and uncoordinated initiatives for the different transport modes. Before 1987, Danish transport policy was particularly concerned with investments in road infrastructure and the desire to build a network of motorways had fluctuated over the years.

Sweden has a longer history of transport policy than Denmark. Preceded by ten years of considerations on the part of an Investigative Commission, the first coherent national transport policy was presented by a Social

Table 8.1 Basic demographic and transport data for Denmark, Sweden and the Netherlands

Country	Population million people	Land area (km^2)	Roads (km)	Railways (km)	Roads (metres per person)	Rail (metres per person)
Denmark	5.333	42,394	71,474	2,859	13.4	0.53
Sweden	8.875	410,934	210,760	12,821	23.7	1.44
Netherlands	15.981	33,883	125,575	2,739	7.8	0.17

Source: *2001 World Fact Book*, Central Intelligence Agency, USA), http://www.cia.gov/cia/publications/factbook

Democrat government in 1963. The policy was accepted unanimously by Parliament in December of the same year. Earlier political decisions on transport and traffic had been less coordinated. The aim of the new policy was to provide the population in different parts of the country with safe and satisfactory transport facilities at low cost, also keeping in mind the social costs of transport. Subsequent deregulation of the transport sector was intended to lead to more competition between different transport modes and to stimulate technical development.

Following the second oil crisis of 1979, a special programme for promoting energy efficiency in transport was formulated. It was based upon the widespread belief that oil-based fuel and oil products would become increasingly scarce. The programme included a strengthening of the role of public transport, stimulation of car-pooling, promotion of energy-efficient vehicles by levying a progressive fuel tax, and various means to stimulate the introduction of new automobile technology. There was also some interest in the development and use of alternative fuels.

The first coherent Netherlands transport policy programme was elaborated in the late 1970s. A 'First Transport Structure Plan' (*Structuurschema Verkeer en Vervoer*) was presented to the Dutch Parliament in 1977 and accepted by Parliament in the session of 1981 (Netherlands, 1977). Congestion was identified as one of the most serious problems of the transport system, due to the ongoing 'explosion of mobility' (*mobiliteitsexplosie*). In spite of this, car ownership figures, at 250 cars per 1000 inhabitants in 1975, were still fairly low in the Netherlands. The solution to problems of congestion was thought to be the construction of more roads, but an increased role for public transport was also indicated as a means of reducing growth in car use. The environment was perceived at the time as a less important issue than congestion and proper provision for transport demand.

Transport policies after 1987

The issue of global warming together with the publication of the Brundtland report stimulated governments to introduce new transport policies around 1990 in Denmark and the Netherlands and, somewhat later, in Sweden. The new era began with a political reconstruction of the perception of the problems of the transport sector. This reconstruction not only sought to account for the risks of global warming but also contained a reinterpretation of the local and regional environmental problems, which now seem to have been taken more seriously than before.

However, Denmark and the Netherlands, on the one hand, and Sweden, on the other exhibited different responses to the new, more alarming view of the environmental problems associated with transportation. As far as their stated transport policies are concerned, the Danes and the Dutch responded to the new situation more rapidly than the Swedes. In 1988 the

Danish Conservative–Centre coalition government published a document that was the first national follow-up of the report of the Brundtland Commission. In this document, a national strategy for the environment and development was presented indicating energy and transport as two societal sectors particularly crucial to the implementation of the strategy. These sectors therefore became the subject of intensified studies. After these studies had been completed, a first Transport Action Plan for Environment and Development was published by the same government in 1990. It was followed up in 1993 by a more detailed Traffic Plan issued by a new government under the guidance of the Social Democrats (Denmark, 1994).

Almost immediately after the publication of the Brundtland report in 1987 'sustainable development' was designated by the Dutch government as the general principle underlying its overall policy (Bressers and Plettenburg, 1997: 125 *et seq.*). In the year 1988, the Dutch coalition government, consisting of Christian Democrats and right-wing Liberals, described the new picture of the problems of the transport sector in 'Part A' of the Second Transport Structure Plan (Netherlands, 1988). Here, the problems of acidification (of soils and vegetation) and of global warming, the latter leading among other things to rising sea levels, were strongly emphasized. Moreover, as Hajer (1995: 194) reports, Queen Beatrix of the Netherlands in her Christmas address to the nation threw her weight behind an apocalyptic vision of environmental decay, adding urgency to the response. After two years, a definite transport plan (*Second Transport Structure Plan, Part d: Government Decision*, 1990) based on this perception of the problems was accepted by the Dutch Parliament in spring 1990.

In Sweden, a new transport policy was presented by the Social Democratic government in 1988, which did not mention the greenhouse effect at all. Nor had the report of the Brundtland Commission any visible impact on Swedish transport policymaking at this time. Somewhat later, in 1991, the still ruling Social Democrat Party presented a bill on environmental policy describing the environmental threats in much more serious terms than before, now including the risks of climate change.

After the elections in September 1991, the new Swedish Conservative-Centre government recorded its support for the transport policy of 1988. The text (Spring 1992) included, however, a more alarming view of the problems of climate change and the more local environmental problems of acidification, eutrophication, photochemical oxidants and air pollution. However, the *final* step toward a more alarming view of the environmental problems of transportation was not taken until March 1998, when a Social Democratic government presented a bill for a new transport policy entitled (in translation) *Transport Policy for Sustainable Development.* This bill did not contain any detailed discussion of the environmental problems of transport but subscribed to the view generally accepted at the Rio Earth Summit of 1992. The Government also stressed the risk of serious conflicts between welfare and economic growth, on the one hand, and, on the other, the necessity of creating a transport system compatible with ecological constraints.

Is it possible to find a theoretical perspective able to explain the policy changes just described? Two theoreticians had, independently and uninfluenced by the Brundtland report, suggested similar explanations for changes in government transport policies. Le Clercq (1987) based his view upon empirical studies of the history of Dutch transport policy, and Starkie (1987) on the history of British transport policy. Both contributions discuss the emergence of new issues in transport policy and the decline of other issues, leading to the modification of government goals and objectives. They are both also concerned about the varying popularity of different policy instruments.

Le Clercq (1987: 93 *et seq.*) thinks that transport policy issues exhibit a pattern of 'upswings' and 'downswings'. Attention paid to a transport issue (for instance, interest in infrastructure investment) fluctuates over time in a cyclical manner. Starkie (1987: 269 *et seq.*) takes a similar view that 'public attention rarely remains focused upon any one issue for very long'. He therefore talks about 'issue cycles', each of which can be divided into five stages of varying duration. He distinguishes between (i) the 'pre-problem' stage; (ii) the stage of 'alarmed discovery and euphoric enthusiasm'; (iii) a stage characterized by understanding of the cost of significant progress; (iv) consequent gradual decline of interest and finally; and (v) the post-problem stage.

Starkie argues that new problems do not necessarily become an 'issue' unless accompanied by some dramatic event, highlighted by the media. The role of interest groups is often crucial in building up pressure. Exposed to intensive and lasting pressure, the old policy collapses. The policy reaction to the built-up pressure depends on various factors such as the character of the issue, current levels of aspiration and the availability of policy instruments adapted to solving the problem. Developments in Denmark and the Netherlands resemble, in my interpretation, the step from Starkie's stage 1 ('pre-problem') with global warming becoming an issue, to stage 2 ('alarmed discovery and euphoric enthusiasm'). In Sweden the step from stage 1 to stage 2 was integrated with stage 3 ('realisation of the cost of significant progress'). In all three countries, however, a new cycle of transport policy was initiated in the 1990s.

This new cycle was marked by the introduction and formulation of a new goal or – to put it in a more theoretical way – of a new 'storyline' called 'sustainability'. Hajer's discourse-theoretic concept of 'storyline' can here be used to illuminate our present account of policy change (see Hajer, 1995: 52–3). This storyline was used either in the narrow sense of 'environmental sustainability' (as in the case of Denmark and the Netherlands) or in the broader sense of 'environmental, economic, social and cultural sustainability' (in the Swedish case). The new storyline was added to the traditional ones in transport policy of efficiency, safety and equity.

There are, after all, significant differences between the three countries in their transition to a new cycle of transport policy. The preparation of a Dutch transport policy for a sustainable society began in 1988 with publication of the draft of a new national Transport Structure Plan (mentioned above). This document aroused great interest in the Netherlands: 'No policy proposals have ever elicited such a massive response. Local authorities, industry and large numbers of social organizations and advisory bodies have expressed support for and criticism of aspects of Part A' (Netherlands, 1990: 5). Many of these actors 'sought to bring maximum influence to bear on decisionmaking at an early stage' (ibid.). The Dutch Government found in 1990 that 'there is every reason to set out a more ambitious programme than was felt necessary in 1988' (ibid.).

In Denmark the existence of a Green majority in Parliament and the activities of a Green lobby seem to have been important precursors of the Danish Transport Action Plan of 1990. The Ministry of Environment may also have played an inspiring role. Certain studies carried out by transport consultants signalled a first step towards the introduction of a Danish expert culture in sustainable transport policy.

The Swedish case again exhibited a different policy procedure. The first indication of the emergence of a new storyline in transport policy came in a bill on road and railway infrastructure investment in 1993. Here, it was said that Sweden had to develop a long-term sustainable transport system in accordance with the agreements entered into at the Rio conference in 1992. In the same year, the Minister of Transport indicated that a 'long-term sustainable transport system' should be created in Sweden. A similar view was repeated when in 1994 a new Investigative Commission was entrusted with the task of preparing a new Swedish transport policy.

To move towards the goal of environmental sustainability, this Commission proposed ambitious quantitative environmental targets and a successive increase in tax on carbon dioxide emissions 'to influence individual and corporate transport choice and behaviour' (Sweden, 1997: 45). The entire proposal of the Commission provoked strong criticism from important actors. In accordance with the Swedish tradition, the report of the Investigative Commission was sent for comment to various institutions, authorities and organized interests. In response, many arguments critical of the report's position were received (see Sweden, Ministry of Transport, 1997). It was afterwards shown that the Swedish Road Federation lobbied very effectively against the proposal.

When the Swedish Government presented its Bill (Sweden, 1997/98: 56) for a new transport policy in March 1998, the overall goal of Swedish transport policy was presented as being based upon the achievement of a transport system that is environmentally, economically, culturally and socially sustainable and, at the same time, efficient in economic terms. This overall goal was translated (in the Bill, p. 17) into a number of partial objectives – not very appropriately, if one pays attention to the overall goal:

- a transport system accessible to all;
- high quality of transport in terms of reliability, regularity and security;
- safe traffic;
- an environment without severe problems; and
- positive regional development.

Most of the partial objectives were in turn provided with time-tabled targets (for instance, for reduced road injuries and reduced emissions of pollutants). The strong economic policy instruments proposed by the Commission were, however, rejected by the Government.

The stated transport policies of the three countries can easily be related to the theoretical concept of 'ecological modernisation' (on this concept see for instance Hajer, 1995; Jänicke and Weidner, 1995). This concept deals with the political capacity of a country to tackle environmental problems built on economic performance, capability for both innovation and consensus-building, and proficiency in strategic policymaking.

In my comparative study of transport policies in Denmark, Sweden and the Netherlands referred to in the introduction I claimed that the most outstanding ecological modernization capacity of the three countries in the 1990s was to be found in the Netherlands (Tengström, 1999: 134 *et seq.*). There are several reasons for this judgement that can be summarized as:

- a rapid political response to new environmental problems resulting from transportation;
- integration of these problems into an intense public debate;
- an elaborated strategy to meet the problems; together with
- innovative thinking in setting objectives, selecting policy instruments and policy evaluation.

The actual outcome of the ecological modernization capacities in the three countries is quite another matter and will be dealt with below. First, however, how are the differences among the three countries in their policy responses to a revised view of the environmental problems of the transport sector to be explained?

Differences among the three countries

The comparatively slow Swedish response to the new picture of the environmental problems of transport is, in my view, primarily explained by the Swedish administrative and political culture. Sweden had developed a long tradition in transport policymaking, particularly in comparison with Denmark but also in comparison with the Netherlands. New transport policies were traditionally preceded by long periods of studies and consultations (extending over quite a few years). In this respect, the role of Investigative Commissions is very significant. Such Commissions include not only representatives of political parties in Parliament but also of affected interests ('the stakeholders') and specialists from administrative agencies. Their task is to produce more knowledge about the issue in question, to promote political consensus, to reduce conflict among organized interests and to offer government the possibility of controlling the formulation of new policies. This institution has been regarded as the Swedish model of capacity building (Lundqvist, 1997: 61).

A second factor which may explain the comparatively slow Swedish reaction to new environmental threats is to be found in the internal debates of the governing Social Democratic Party in the late 1980s. In its bill on environmental policy of March 1988, the Social Democrat government emphasized, at the rhetorical level, the necessity of respecting the limits of nature by means of new lifestyles and new patterns of consumption (as a concession to the Green opposition within the party). In reality, however, there was no deviation from the traditional attitude of the majority. This attitude was based on the view that environmental problems of production generally and transport in particular could and should be solved by means of developments in technology (Anshelm, 1995: 120–4).

The Dutch political context was different. In the late 1980s, the Dutch Government was inspired to increase its commitment to environmental objectives by the report of the Brundtland Commission (1987). An important role was also played by the Minister of the Environment personally (Bressers and Plettenburg, 1997: 124 *et seq.*) and by some alarming reports of the National Institute of Public Health and Environmental Protection (*op. cit.*, 121 *et seq.*). When the Government initiated a public debate on transport policy, both the general public and non-government organizations were immediately confronted with new aspects of the environmental problems of transportation. This process was characterized by open conflict and debate in contrast to the conflict-reducing strategies typical of Denmark (see below) and Sweden (see above). In 1990, the Second Transport Structure Plan was accepted by the Dutch Parliament after some modifications initiated by the public debate.

There may also be geographical reasons for the rapid reaction of Dutch opinion to the risks of climate change. Global warming caused by an enhanced greenhouse effect posed a special threat to the Netherlands, one half of whose land area is already below sea level. The link between the greenhouse effect and rising sea level was made plain in the original of the Second Transport Structure Plan (Netherlands, 1988: 11).

The political situation in Denmark differed from that of Sweden and the Netherlands. As Christiansen (1996: 89) points out, since the 1970s, Danish environmental policy had 'mobilised environmental groups, business organizations, and political parties, as well as individual politicians building a political platform on the environment as a political problem, but the political and administrative responses were shaped within the traditions of co-operation and consensus so characteristic of Danish political life'. The existence of a Green majority in Parliament was probably the decisive factor in the treatment of environmental issues. The contemporary Danish political situation can therefore be regarded as one of the key factors underlying the rapid Danish response to a more alarming view of the environmental problems of transportation. Also, in my view, a certain role was played by the fact that, in contrast to Sweden, transport policy had not yet become institutionalized practice in Denmark.

The different roles of the Green movements in the three countries may also have contributed to the differences in the transition. These movements have very specific characteristics. Danish environmentalism has been associated with 'the populist "grass-roots" tradition of Danish political culture' (Jamison *et al.*, 1990: 192). This feature may have helped prepare the political ground for early Danish acceptance of a new view of the environmental problems of transport. In the Netherlands Green movements were more professional and usually mobilized voters by means of expert-orientated and 'legalistic' arguments (ibid.: 187). Environmental organizations in Sweden can be seen as more 'technocratic', and the Swedish attitude to environmental problems appears to be mainly pragmatic and utilitarian (ibid.: 188).

To summarize: the differences between Danish, Dutch and Swedish political responses can primarily be ascribed to differences in political and administrative culture and, in the Danish case, to the prevailing political situation. The differences may also be associated with the role of Green actors in the Danish case, and with geographical factors in the Dutch case.

The outcomes of the new transport policies

In this section, the real outcomes of Danish, Dutch and Swedish transport policies in the period 1987–97 will be analysed with special reference to environmental sustainability. For a complete set of tables with detailed figures I refer to my book mentioned in the introduction to this chapter (Tengström, 1999).

In the Danish, Dutch and Swedish political documents of the period 1987–97, it is possible to identify some intermediate objectives which may be interpreted as steps towards environmental sustainability in transportation. These comprised a political will to influence:

- transport volumes (in terms of size and/or distribution among different transport modes);
- the level and/or composition of energy consumption;
- the technical standards of the motor vehicle fleet;
- the environmental adaptation of new infrastructure.

The outcome of these political ambitions can, in most cases, be evaluated by quantitative methods, as time-tabled quantitative targets for the reduction of emissions from traffic were explicitly defined. There are, however, a number of problems associated with the measurements of the real outcome of policies. The first has to do with the availability of quantitative data and with their reliability. Secondly, the international comparability of the figures describing the emissions is also sometimes problematic. Finally, one cannot regard as self-evident figures describing broken trends as being results of transport policy and not of other factors. For reasons of space I shall here confine myself to just a few tables. In the first place a success story will be told. After that, some examples of political failures will be demonstrated.

All three countries have striven for a reduction of the emissions of NO_x from road traffic. These efforts were all fairly successful in the years between 1986 and 1995 as can be seen from Table 8.2.

The rapid reduction of emissions in Sweden may be questioned and the figure must be characterized as relatively uncertain (plus or minus 10 per cent – see Tengström, 1999: 183).

Now let us turn to the political failures. The increase in passenger kilometres travelled by car (by drivers and passengers) has in most cases been greater than the increase in gross domestic product (GDP). This development was not in line with policy (see above). The Netherlands was, to some extent, an exception (Table 8.3).

Many politicians hoped that technical development would reduce fuel consumption in petrol-driven cars. There is, in fact, a weak increase in the number of kilometres driven per litre of fuel in the early 1990s in all three countries but this trend was soon broken (Table 8.4).

Stabilization of the emissions of CO_2 has also been a declared political target in Denmark, the Netherlands and Sweden. No such stabilization was in sight between 1986 and 1995 (Table 8.5).

Table 8.2 The development of emissions of NO_x from road traffic in the three countries for four selected years (index 1986 = 100)

Country	1986	1990	1993	1995
Denmark	100	105	95	86.5
Netherlands	100	101	96	91
Sweden	100	97	81	77

Sources: Denmark: *Natur og Miljø*, 1997 (with a personal communication). Holland: *Beleidseffectrapportage*, 1993, pp. 38, and 1995, p. 52. Sweden: *Environmental Report*, 1996, ed. by the Swedish National Road Administration, p. 12 (together with a personal communication).

Table 8.3 The development of passenger kilometres by car (drivers and passengers) in the three countries compared with the development of GDP (index 1986 = 100)

Country	1986		1990		1992		1995	
	GDP	km	GDP	km	GDP	km	GDP	km
Denmark	100	100	104	115	106	120	114	130
Netherlands	100	100	113	107	117	110	125	118
Sweden	100	100	108	115	106	117	112	121

Sources: GDP: *Statistisk Årbog*, 1995, p. 526; 1993, p. 526 and 1997, p. 521, ed. by Denmarks statistik. Transport: Denmark: *Natur og Miljø* 1997 (with a personal communication). The figure of 1995 is estimated. Netherlands: *Beleidseffectrapportage 1993*, p. 60 and *1995*, p. 74. Sweden: *Transportprognos år 2005 och 2020*, ed. by the National Road Administration.

Table 8.4 The average energy efficiency of new petrol-driven cars measured as number of kilometers driven per litre of fuel (index 1985 = 100)

Country	1985	1992	1994	1995
Denmark	100	108	104	n.a.
Netherlands	100	100	102	92
Sweden	100	104	101	102

Sources: Denmark and Sweden: *Transportation Energy Data Book Ed. 17*, (table I 8). Netherlands: CBS, Voorburg/Heerlen.

Table 8.5 Emissions of CO_2 from road traffic in the three countries for four selected years (index 1986 = 100)

Country	1986	1990	1992	1995
Denmark	100	106	108	114
Netherlands	100	112	117	123
Sweden	100	106	109	108

Sources: Denmark: *Natur og Miljø* 1997 (with a personal communication). Holland: *Beleidseffectrapportage* 1993, p. 42 and 1995, p. 56. Sweden: *Environmental Report* 1996, ed. by the Swedish National Road Administration, p. 12 (together with a personal communication).

As to infrastructure, there was a substantial increase in the length of motorways in the first half of the 1990s in Denmark and Sweden (Table 8.6).

Expansion of the motorways continues a historical trend. In Denmark the increase had been 200 per cent between 1970 and 1989. In Holland and Sweden there was an increase of 111 per cent and 132 per cent respectively in the same period.

Table 8.6 The expansion of motorways in the three countries between 1990 and 1994 in absolute (km) and relative (%) numbers

Country	1990 (km)	1994 (km)	Change (km)	Change (%)
Denmark	604	786	182	30.0
Netherlands	2092	2167	75	3.5
Sweden	929	1061	132	14.0

Sources: *World Road Statistics* and, for Sweden, the National Road Administration.

The evaluation of Danish, Netherlands and Swedish transport policies in the previous section is based on data that have to be interpreted with great caution. However, there is one undeniable success story: the reduction of the emissions of NO_x. This success is, of course, associated with the spreading of cars equipped with catalytic converters in accordance with political decisions at both the national and the European level. The original cause of this success is not to be found in the national policies of the three countries, nor in the transport policy of the European Union. It was rather a matter of 'technology forcing' by the US federal government that compelled the international automotive industry to develop some technical device to prevent the negative impact of the emissions of NO_x and some other substances. Only after that did the European nations introduce the mandatory fitting of catalytic converters but at a much later date than the USA and Japan.

The Netherlands' attempt to initiate a certain decoupling of economic growth and increased individual mobility, above all by car, can also be counted a kind of success (see Table 8.3). This success had probably to do with the national transport policy of 1990. It was, however, only a partial success. The problems of increasing car traffic are still a source of great concern in the Netherlands.

The period 1987–97 was also marked by serious transport policy *failures* in the perspective of environmental sustainability (for full evidence see Tengström, 1999). The failures can be related to the four intermediate objectives identified above:

1. Despite the political will to influence transport volumes in terms of size and/or distribution among different transport modes: (i) transport volumes were still increasing (somewhat less rapidly in the Netherlands); (ii) car density was still increasing (with the temporary exception of Sweden, as an effect of weak economic development in the first half of the 1990s); (iii) passenger kilometres by car were still increasing; (iv) the share of public transport versus private transport was not strengthened; and (v) the role of the bicycle was not strengthened.

2. Despite the political will to influence the energy consumption of the transport sector: (i) the use of energy in the transport sector showed a stable upward trend (with the temporary exception of Sweden between 1996 and 1997); (ii) the dominance of fossil fuels was still unbroken; and (iii) the per capita emissions of CO_2 were far from acceptable from a global perspective.
3. Despite the political will to influence the technical standard of the fleet of motorcars: (i) the trend towards improved energy efficiency was broken during the period; (ii) the percentage of heavier cars increased; and (iii) the emissions of CO_2 from the transport sector increased significantly.
4. Despite the political will to influence the environmental adaptation of new infrastructure: (i) losses of productive soil were still substantial as an effect of the building of new motorways (to a somewhat lesser degree in the Netherlands where the total losses are, nevertheless, impressive) and (ii) the attempts to apply environmental impact assessment in road building were a failure.

How to explain the failures?

Economists traditionally explain the problems of the transport sector in terms of 'market and government failures'. An example of this is the report entitled *Market and Government Failures in Environment Management: the Case of Transport* (OECD, 1992). The 'market failure' explanation is based on the view that the external costs of traffic (pollution, noise, accidents, congestion, etc.) are not internalized and not paid in full. Many economists – and environmentalists – therefore recommend governments to increase taxes on transport. If they do not follow this recommendation, they are seen as responsible for a 'government failure'.

My approach to the concept of 'government failure' is different. I look upon 'government failures' in the perspective of political science rather than in that of political economy. I believe that the present government failures can be explained by certain shortcomings in national policymaking. I claim that the politicians in the three countries have failed:

- to analyse the inherent conflicts between the new goal of environmental sustainability and traditional goals in transport policy (Netherlands politicians may be an exception here);
- to consider the possibility that the attainment of the new goal requires quite a new package of policy instruments; and
- to make a realistic analysis of the problems of implementation of policy instruments supporting the goal of environmental sustainability.

All these shortcomings are evident at the national level. Environmental and transport policy, however, is also being dealt with within the European Community as a whole. National politicians, therefore, face a difficult situation. On the one hand, they cannot solve the problems of present unsustainable transport systems exclusively at the national level, because: (i) the environmental problems are not confined within the frontiers of the nation; and (ii) a national environmental decision may be interpreted by the Commission as introducing a new trade barrier. On the other hand, any efforts made by national politicians to realize their transport policy goals at the European level are circumscribed by the possibility that a proposal concerning a harmonized transport policy may be rejected (with reference to the principle of subsidiarity) or a decision about harmonization of rules and fees may not receive unanimous support.

Therefore, I draw the conclusion that, in the present European context, national politicians in the member states of the Union are caught in a kind of a social trap in their efforts to create sustainable transport systems. To some extent, the present institutional structures of the European Union have to be blamed for this situation.

What of the role of major collective actors? There are a number of important actors besides the national governments that are able to influence the expansion of the transport infrastructure, the composition of the vehicle fleets, the consumption of fuels (type, quality and quantity) and the development of individual mobility patterns. Such actors are governmental agencies, local political bodies, producing industries, automotive industries, transport companies, tourist organizations, producers and distributors of fuels, trade unions and other non-governmental organizations.

The collective actors' own perceptions of the problems and possibilities of the present transport systems are important realities. There are (to my knowledge) no empirical studies of how different collective actors actually perceive the problems of the present unsustainable character of the national transport systems. However, it may be assumed that there is little disagreement about the goal of sustainable transport but a host of diverging ideas about when and how to define this goal and how to realize it.

The roots of these diverging ideas are to be found in various *social representations* of the problems and possibilities of the present transport systems (on the concept of 'social representation' see Moscovici, 1989). These views have been shaped (through omissions, additions and perversion) in a way that satisfies the needs and interests of the different collective actors. The interactions among these actors and between them and their government are influenced by the differences in their respective interpretations of the real world.

Hajer (1995) would probably analyse such representations that dominate in terms of *story lines*. As a consequence of discourse struggles based on different story lines, important actors can therefore be said to be unable to co-operate in the national interest of reducing the unsustainable character of the transport system and its threats to the climate system and to the function of terrestrial and maritime ecosystems. The result is often 'interaction failures' (a concept introduced by Tengström *et al.*, 1995). Today no collective actor is strong enough to impose its own interpretation of the problems and possibilities of the transport system (its own story line) against the will of the other actors.

What of the role of ordinary citizens? In their capacity as individual transport users citizens determine, to a large extent, the number of trips undertaken and the modal split. Their vehicle preferences influence the real outcome of any political attempt to reduce the negative environmental impact of transportation. In their role of voters they also define the action potential of the policymakers. Thus, as consumers and voters, citizens contribute substantially to the failures of transport policy.

First, it can be assumed that citizens are most concerned with the practical problems of their own everyday transport. Therefore, they spend very little time on thoughts about the long-term problems of the national or European transport system, even if many of them might be aware of the existence of such problems. Their attitude can be understood in the perspective of the German philosopher Jürgen Habermas (1981). Citizens are mostly concerned with the problems of their own *life-world* and very little concerned with the long term problems of *systems* such as the transport system, the climate system or the terrestrial and maritime ecosystems. This 'system-world' is mostly distant, while their 'life-world' is ever present. It can be concluded that, in their eyes, problems of increasing costs, poor efficiency, deficient safety and remaining inequalities (for instance, between the sexes and between various income groups) take priority over the problems of long-term unsustainability of the transport systems. Secondly, unimpressed by new and alarming statements from transport experts, most ordinary citizens do not approve the use of ecologically efficient instruments in transport policy, for instance substantially increased fuel taxes. This situation is aggravated by the fact that the present use of public information on the part of the government is generally regarded as inappropriate, and the results of public information campaigns have mostly been insignificant.

These circumstances explain the attitudes of the citizens that result in what I would like to call *acceptance failures*. This factor aggravates the other forces influencing national politicians in their attempts to improve the environmental sustainability of transportation. At the same time, there are probably quite a few citizens that really have modified their views of the problems and possibilities of the transport system. Their changed perceptions do not lead, however, to any behavioural change in transport matters. Many individual citizens find it pointless to alter their behaviour as long as they do not believe that a substantial number of other citizens will do the same: this can be seen as an example of the philosophical problem of 'collective action' or as an example of the well-known sociological theory of 'social dilemmas'.

Besides this, some citizens may have very ambivalent attitudes to car use. They like to drive their cars but they are also aware of the negative environmental impact of mass automobility. According to the psychological theory of 'cognitive dissonance', they tend to belittle the negative aspect of car use – for as long as they can.

It should therefore be recognized that acceptance failure has something to do with the cultural role of the automobile in Westernized countries. In the perspective of social anthropology, the motor car is a well-known carrier of cultural values such as freedom, status, wealth, togetherness, masculinity, etc. It is therefore understandable that transport policies questioning the present expansion of car use appeal neither to current nor to potential car users. The problems of the currently unsustainable transport systems in countries such as Denmark, the Netherlands and Sweden expose their citizens to questions of a deep existential nature concerning their relation to an artefact called 'the motorcar'. The answers to these existential questions will be crucial for the future prospects for transport policies aiming at the creation of long-term sustainable transport systems.

Prospects for the future

What will happen in the future? The creation of long-term sustainable passenger transport systems necessitates, in my view, the involvement of the ordinary citizens. To achieve their involvement, traditional policy instruments seem to be inappropriate. More importance needs to be attached to an underestimated policy instrument, namely *communication*.

Communication is a complex interaction between, at least, two participants. If distinguished at all from merely informing, the official view of communication as a political instrument is often technocratic. Ordinary citizens are seen as objects of educational efforts rather than subjects equipped with intellectual capacity of their own. I believe that policymakers and their experts have to change this view, if they want ordinary citizens to become more involved in the necessary transformation of the transport system.

There are a number of well-known difficulties associated with the use of communication as a political instrument. Communication is a time-consuming process even when successful. It is easier to initiate communication at the local level than at the national level. The channels between the politicians and the general public are some times non-existent, sometimes difficult to use.

The entire idea of the involvement of ordinary citizens in transport policy can be linked to Giddens' theory of 'life-politics' (Giddens, 1991). According to him, the global problems of environmental sustainability necessitate not only coordinated global responses but also 'reaction and adaptation on the part of every individual' (222). With reference to these ideas the creation of a sustainable society necessitates the involvement of the ordinary citizens, and maybe also their deep commitment. The sustainability transition has to become a democratic process (see O'Riordan, 1996). Most decisions concerning whether, when and how to move from one place to another are taken individually, but in the light of that fact the transformation of the passenger transport system could be seen as a process of collective learning.

In the case of environmentally sustainable transport, there is possibly general agreement about the necessity to develop transport systems that are sustainable in the long term, but ideas diverge about how and when to reach this goal. Conflicts between collective actors may, however, become constructive, particularly if these collective actors consist of reflecting citizens. My main thesis, therefore, is that there is reason to believe that only a period of intense communication and conflict initiated by the policy-makers will change the preconditions for more radical policy options.

9

The Privatization of the Japan National Railways: the Myth of Neo-Liberal Reform and Spatial Configurations of the Rail Network in Japan – a View from Critical Geography[1]

Izumi Takeda and Fujio Mizuoka

Introduction

Successful reform of the Japan National Railways (JNR) has been touted as a model of deregulation and privatization. However, close scrutiny of the process and consequences gives us a picture quite different from this imagery. This chapter analyses the changing configuration of the national railway network of Japan as the outcome of the privatization of the Japanese National Railways (JNR hereafter) into seven JR companies – six regional passenger companies and a nationwide freight company – which took place on 1 April 1987 (Figures 9.1a and b).

Development of the railway network in Japan before the Second World War

At the beginning of the Meiji era when the modernization of Japan started, the national government did not have sufficient funds to create a nationwide rail network. Yet from 1880 onwards construction of railways received top priority in government policies. In order to accelerate construction, the government adopted the recommendation of a British consultant to use the gauge of 1.067 metres instead of the standard 1.435 metres. To compensate for the lack of government funds, private railway companies were called in to construct some trunk lines necessary for the nationwide rail network. These private lines were later nationalized through the Railways Nationalization Act of 1906, which reduced fares for freight transportation by reducing fares for

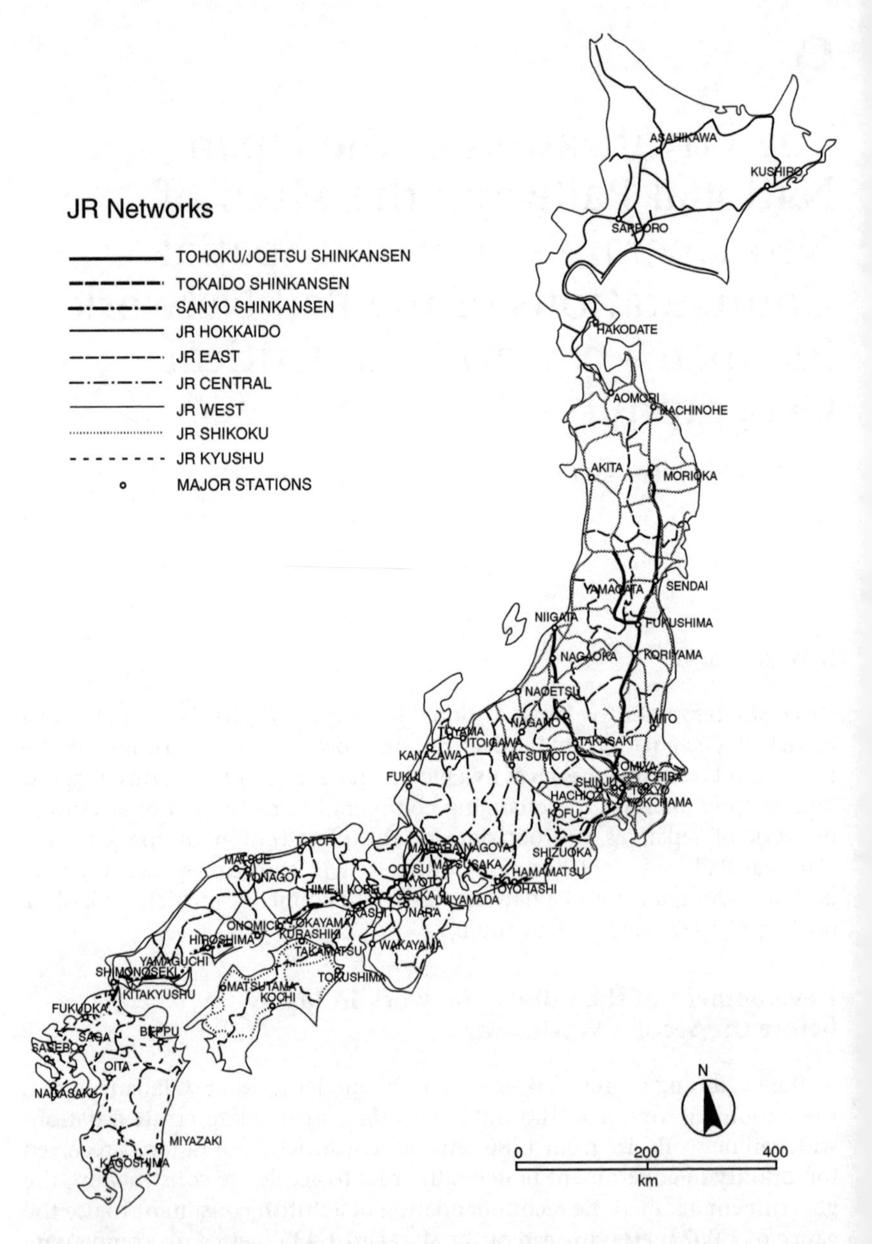

Figure 9.1a The present JR network with divisions into six regional passenger companies. (Map redrawn from original JR map by Chandra Jayasuriya.)

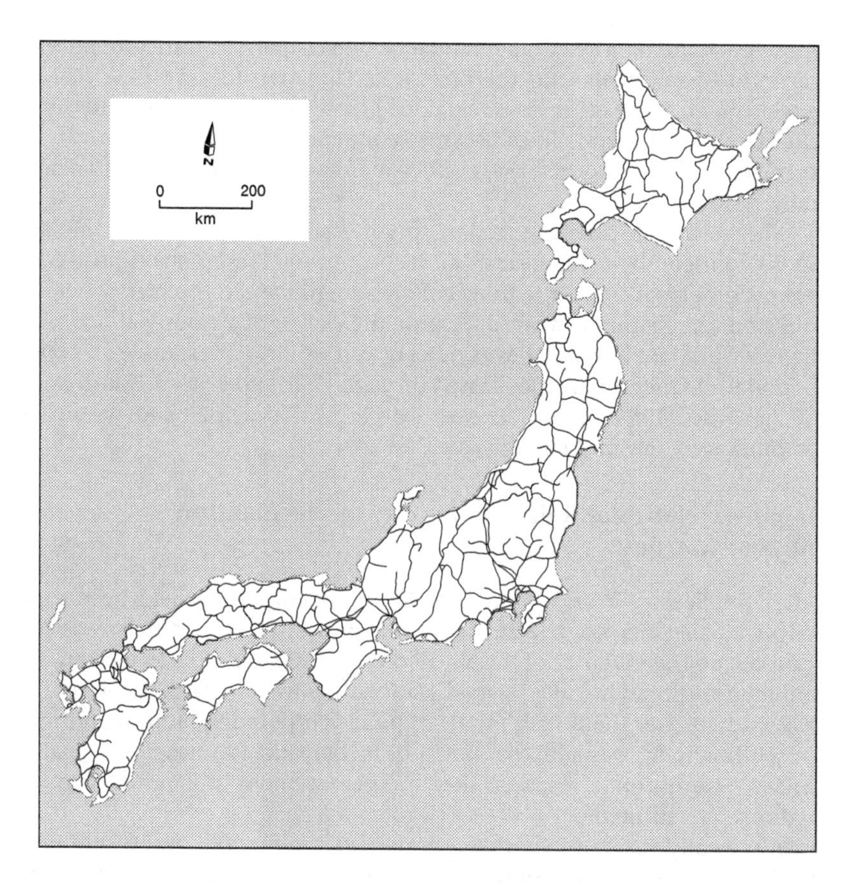

Figure 9.1b The network of the former Japan National Railways at its densest in the early 1980s

longer distance trips. The nationalization was also regarded as necessary from the military standpoint.

In the 1910s and early 1920s, the emphasis of railway construction shifted from trunk lines to the local branch lines. The Railways Construction Act, amended in 1922, drafted the eventual shape of the nationwide railway network, with actual routings specified. The instigation for construction of these branch lines came from parochial demand made by local class alliances led by local men of high repute, and was materialized through the Seiyu-Kai (a political party in pre-War Japan) political machine.

Even after the nationalization of the trunk lines, construction of railways by private concerns continued until the Great Depression, driven by speculative motives. Some railway companies had suffered financial difficulty once the operation had begun, and the lines were later transferred to the national government. Some other lines remained private until this point, and the territories served by these lines became economic fiefdoms of the private rail companies. These processes gave rise to a 'homogeneous space' with a dense railway network.

With respect to the connection to continental Asia, a nodal hub-and-spoke configuration was clear from the beginning. Prestigious express trains connecting Japan proper with its colonies originated in the piers where ferries from the mainland arrived, passing the capital of a colony at odd hours to arrive in Manchurian cities as quickly as possible. For example an express bound for Manchurian cities passed through Keijo (now Seoul, South Korea) at 3am (Ko, 1999). Train services catering to local demand within the colonies were few and far between.

Japanese National Railways and its modernization in post-war days

After the Second World War, the railway system, which had been under direct management of the Ministry of Railways in pre-war days, was reorganised, on the initiative of the allied forces (GHQ) that occupied Japan, into a public corporation called 'Japan National Railways' (JNR). Although JNR was created after the British model of public corporation, and the principle of self-financing was adopted, much of its bureaucratic structure remained intact. The ultimate responsibility for management still rested with the national government.

In the early years of JNR, the railway held a monopoly in the land transport market, and it was simple to earn massive revenue from trunk lines which was then used to cross-subsidise the local branch lines. However, the monopoly was gradually eroded by automobiles and airlines. The need to maintain the network in rural areas and improve it in urban areas in order to provide better a railway service and meet challenges from the parallel private lines imposed a substantial financial burden. JNR set up its own investment projects for electrification and double-tracking, eliminating steam locomotives and introducing newly built rolling stock, and implemented these projects out of its own revenue and debt-financing.

The implementation of massive modernization plans and maintenance of cross subsidies helped to keep the dense, relatively homogeneous and efficient nationwide rail network viable. The trunk and sub-trunk lines were served by 'L-express' trains running frequently with identical time intervals, while numerous rail-car expresses were operated on rural lines to connect even the smaller villages to the nationwide rail network in a very convenient way. Furthermore, the Japan Railways Construction Corporation, a government body, was established to continue building new rural rail lines according to the Railway Construction Act of 1922, which had always been demanded by local MPs. The nationwide rail network thus became denser and more homogeneous than ever.

The privatization of the Japanese National Railways

The 30 trillion yen (more than US$250 billion) of debt generated mainly through debt-financing of improvement projects, as well as deteriorating labour relations, made the reform of JNR an inevitable policy agenda in the 1980s. The Board for Administrative Reform, a consultation body of the Japanese government, attempted to solve the problem of JNR radically by privatising the whole organization. This was an enormous political challenge as it involved thoroughgoing organizational reform of the entire railway system. But the government finally privatized JNR in April 1987. The privatization was characterized by six main features (see Kato and Sando, 1973; Mitsuzuka 1974; Kakumoto, 1996).

First, the process was strongly influenced by the *worldwide tendency towards privatization and deregulation*, which drew upon neo-liberal philosophy and neo-classical economic theory. The neo-liberal philosophy, which has increasingly become popular in countries with an Anglo-American heritage, was introduced into Japan in the 1980s in order to achieve 'greater efficiency' in the public services. The Board for Administrative Reform adopted a policy combination of 'geographical split of the system', 'privatization', and a segmented accounting system, thus eliminating cross-subsidies that had formerly blurred the accounting parity across the nationwide system. The expectation was that efficient management of the three companies on Honshu would generate enough profit to repay the accumulated debt of the JNR.

Secondly, the system was disintegrated *horizontally* into six regional passenger companies, which were to keep the fixed facilities under their ownership. This compares with railway reform in Europe, where *vertical* disintegration splits the proprietorship of tracks and other fixed railway facilities from that of rolling stock, operation and marketing of the railway transportation services, in order to bring about competition in railway operation on a par with that taking place in the road or air transportation systems. The actual boundaries between the JR companies on Honshu followed the boundaries of former regional administrative bureaus of the JNR. The resulting geographical boundaries of JR Companies did not necessarily match the prefectural jurisdictions or the action spaces of commuters of some metropolitan areas. For example, all three Honshu JR companies are represented in the single prefecture of Nagano in central Japan. The commuters in Kofu, the prefectural capital of Yamanashi, are forced to transfer from JR East to JR Central every day as they commute from home in the eastern part of the metropolis to offices or schools in the southern part. The JR Freight company was supported by the Management Stability Fund and the railway track usage fee was abated through application of the 'avoidable cost' principle in which JR Freight is charged for the potential savings of the JR passenger companies that would have arisen if JR Freight did not operate freight services using the railway tracks (Hori, 2000).

Thirdly, the split was carried out on the basis of *a mixed principle of physical and financial considerations*. Boundaries were, in the first instance, drawn according to the physical geography of the Japanese archipelago: between Honshu (the main island) and three smaller islands of Japan (Hokkaido, Shikoku, and Kyushu). Arrangements were made to support JR Hokkaido, JR Shikoku and JR Kyushu with the Management Stabilization Fund, in order to reinforce the weaker financial position of these companies due to sparser or smaller population in their market areas. On the main island of Honshu, each of the three major metropolitan areas with high demands for passenger traffic is allocated to each Honshu company: Tokyo – Yokohama to JR East, Nagoya to JR Central and Kyoto–Osaka–Kobe to JR West. The lucrative Shinkansen (the 'bullet train'), built to run on the 1.435-metre standard gauge lines, was also split into three sections in order to provide each privatized company on Honshu with a stable financial base. JR companies in Honshu were expected to cross-subsidise less profitable conventional lines with profits earned from Shinkansen lines and from the metropolitan sectors. The proprietorship of the Shinkansen had been given in the beginning to the Shinkansen Railway Proprietary Agency, from which three passenger companies on Honshu were to lease out the fixed facilities with payment of a fee. Four and a half years later, in a move to straighten the way towards listing their stocks on the Stock Exchange, these Honshu companies opted to purchase Shinkansen property from the Agency at its book value.

Fourthly, the metropolitan cities of Japan have had many private rail lines, which have kept challenging JNR in the past, and have sometimes pushed JNR into a corner. The companies operating these lines solidified their financial position through earnings from property development projects and related businesses operating along their lines, and eventually came to dominate a particular geographical sector of a metropolis, just like a feudal warlord. Such a company would strive hard to uplift the image of the sector under its domination and promote its quality as a residential neighbourhood. This practice of private railway business gave much insight to the execution of the privatisation of JNR, and *the privatized JR companies were asked to follow the past practice of private railway companies*.

Fifthly, each passenger company was nevertheless expected to resort to the *principle of cross subsidy to sustain rural lines within its territory of operation*. Thus, in order to minimize the funds to be transferred to these rural lines, the JR companies increased their drive for 'efficient' and 'rational' operation by classifying lines into two categories according to their profitability.

For the lines with profitable prospects, further investment was made and the service expanded on lucrative trunk and urban routes, where sleek express trains travel swiftly, whereas for money-losing rural sectors deliberate disinvestment and reduced services became the norm. For the lines with little prospect of profits, cheaply built and less comfortable urban-type carriages were introduced into rural sectors, with less frequent services than before. According to Sone (2002), such deterioration of service became more apparent five years after the reform (see Figures 9.2a and b).

Sixthly, the operation of privatized JR companies was infested with the problems inherent in Japanese bureaucracy, such as inconsistency in authority of supervision and lack of external mechanisms of evaluation, in such a way as to reflect little in the way of user needs. As private companies, the JR system is bound to the legal requirement to become accountable to company shareholders rather than the users. The conservative Liberal Democratic Party, in propagating the 'success' of the privatization of JNR, blames the trade unions for the bad management of JNR; and public scrutiny of the privatization was cunningly circumvented. With the absence of the formal procedures for public assessment and evaluation of the operation of JR companies, the *public power to control JR operation has today been considerably reduced.*

Figure 9.2a A more comfortable carriage built by JNR for a local commuter train

Figure 9.2b A new commuter carriage built by JR. Drinking on the train back home was once a common pastime after a hard day's work – but now the new carriage makes it no less awkward

The deteriorating rural sectors and stronger nodality in the rail network under the privatized JR companies

In the process of privatization, the authorities boasted that the railways of Japan would revive and enter a new era of development. The reality was, however, that the 'new era' meant a reduced role of the railways in transport within the nation and a sparser and more nodal spatial configuration of the rail network in Japan. In what follows we examine four features of the new privatized system: the change in spatial configuration from a dense, homogeneous network to a nodal one with large metropolitan cities at the hub, the spatial fragmentation of the fare system, increasingly fragmented train operations and differentiated quality of service, and the emergence of rival, and less sustainable, modes of transportation.

Change in spatial configuration – from a dense, homogeneous network to a nodal one

The Shinkansen has played a major role in transforming the railway network of Japan. Based on the scheme and technology that JNR had been nursing over the years since pre-war times, Japan indeed took the initiative in opening up the era of high-speed trains. The first section of Shinkansen, between Tokyo and Shin Osaka and put into operation in 1964, brought the train into the global limelight, and many parts of the world followed suit, including the TGV in France and ICE in Germany.

Thanks to its capability, efficiency and sheer speed in annihilating space, the popularity of Shinkansen attracted demands from local MPs to build more lines in areas with less population density, just as in the case of conventional branch lines decades ago. The Shinkansen lines have been extended westward to Hakata and northward to Morioka and Niigata. Yet the huge construction costs and the lack of profitability expected from them aroused concerns about further expansion of the Shinkansen system, especially in view of the current huge government fiscal deficit. A concessionary attempt was therefore made to introduce a 'mini-Shinkansen', a dwarfed version built simply by widening the gauge of conventional tracks to 1.435 metres to make through operation of carriages to and from Shinkansen lines possible, or by building 1.067 metre gauge tracks to Shinkansen standard to make high-speed operation (160 km/hour) of express trains possible.

With several lines built already, the Shinkansen network now forms the main artery of the nationwide rail network. In the JR East area, the artery has come to form a clear 'hub-and-spoke structure' with Tokyo as the hub. Those prefectures excluded from Shinkansen have been trying hard to entice the service to their jurisdictions. In the meantime the JR companies discontinued operation of many railcar expresses connecting small and medium-sized towns to each other. As a result, just over a decade after the privatisation of JNR, the spatial configuration of the railway network in Japan has shifted from the 'dense and uniform pattern covering the entire nation' to a 'nodal hub-and-spoke pattern with lines stretching out from large metropolitan cities'.

Spatial fragmentation of the JR fare system

The privatised JR system has been using the same fare table that had formerly been used by the JNR. JR companies deserve praise in this respect. In fact, the JR companies in Honshu have not increased fares at all, apart from passing on the newly-introduced consumption tax, for 14 years ever since privatization. Nevertheless, each JR company now attempts to set up its own fare table, to rectify differences in the financial positions of each JR company arising from differences in the geographical endowments of their respective operational territories. Since the rule agreed at the time of JNR's privatization prohibits spatial disintegration of the nationwide fare system by the JR companies, intense 'local adjustments' have been made to the integrated nationwide fare table.

For sectors where challenges from competing modes of transportation are harsh, discounted fares have been introduced; while on lines managed by less profitable JR companies on the smaller islands, additional fares have been added on top of the standard nationwide fare table. Additional fares are also collected from the sectors in order rapidly to recoup recent investments. Some of the nationwide fare discount scheme such as the *shuyu-ken* (excursion ticket) introduced before privatization were abolished because of the difficulty of allocating revenues from its sales to each JR company. The integrated and uniform nationwide fare system carried over from the JNR has thus come to be very much differentiated spatially to reflect local differences in market potential.

Increasingly fragmented train operation and differentiated quality of service

The shift of the spatial configuration of the Japanese railway system from the 'dense and homogeneous pattern covering the entire nation' to 'nodal and hub-and-spoke pattern with sparse Shinkansen network forming the main artery' has had an effect on the pattern of train operations as well, causing more inconvenience to train travellers.

The fervent desire for construction of Shinkansen lines is still at large in localities where Shinkansen lines have yet to be built. In order to reconcile the local desires with neo-liberalism, the government set up a new package scheme, without enactment, in which the management of conventional lines parallel with Shinkansen should be separated from JR. Their operation is to be placed under the new management of a public and private partnership parallel with Shinkansen fare structure detached from the nationwide system. The conventional lines thus detached cater to local services only, with limited revenue and management bases. The newly created public–private partnerships are thus compelled to raise fares in order to make their finance structure break even. If they want Shinkansen to be built, local governments along the proposed Shinkansen lines have no option but to accept this deal.

The most controversial case is the proposed detachment from JR East of the northernmost section of Tohoku Line between Morioka and Aomori. This section is currently the artery for freight services connecting Hokkaido with the Tokyo metropolitan region. Its importance had increased after the opening of the world's longest submarine railway tunnel under Tsugaru Strait, which physically connected the JNR network in Hokkaido with that in Honshu. The Prefecture of Aomori, which has demanded the extension of the Shinkansen line into its own jurisdiction for decades, has already set up a public–private partnership to manage this section. Things are still very uncertain as to how freight trains on this section can be operated after the detachment. The local citizens of Hokkaido may eventually be asked to pay a higher local price for the commodities in order to defray the increased freight cost charged by the public–private partnership.

For the conventional lines that remain under JR management, direct services running conventional trunk trains have been either reduced or discontinued and split into small sections. A number of overnight services connecting Tokyo and Kyushu have been abolished, as were other long-distance services travelling along the Japan Sea coast. The train services running across the boundaries of privatized JR companies have also been reduced and the introduction of new carriages has been put off, as the operation of 'inter-JR' services now involves burdensome coordination and strife among the independent JR companies.

The local variations arising from differences in market situation is another factor leading to fragmentation of services. Passengers can travel sections where rival operators adopt aggressive strategies on comfortable, newly built carriages for discounted fares, whereas for sections where JR enjoys spatial monopoly they are forced to travel in cheaply built or second-hand carriages for higher fares. Some sections of conventional lines have been physically fragmented, due to widening of the gauge of conventional lines into the standard gauge to create 'mini-Shinkansen' lines. As carriages can no longer make through operation from conventional 1.067 metre to 1.435 metre gauges, passengers are forced to transfer at stations where gauges change.

Emergence of rival modes of transportation

Due to heavier investment in roads supported financially by fuel taxes specially designed to channel investment to their construction, automobiles and buses have been gaining a stronger competitive edge over rail (Nakanishi, 1985; Kamioka, 1994, see also Chapter 15 of this volume. In 1960 there were only 16 cars per 1,000 people). The neo-liberal policy requiring segmented accounting has caused branch rail lines in sparsely populated areas to lose out. In such areas it is becoming normal for the facilities to be left to deteriorate without repair, and for services to be reduced with longer connection times and occasional suspension for maintenance. The people are forced to use other modes of transportation than railways in travelling in the rural areas. The JR companies themselves have recently become more eager to promote rental cars rather than local branch lines for rural tourist traffic in conjunction with Shinkansen (Figure 9.3).

The future does not seem bright also for trunk lines of 1,000 kilometres or longer. Projects to construct airports have taken place in many regional centres and airlines are aggressively competing with long-distance rail services. The whole process reflects hostile relations between the Ministry of Construction and the Ministry of Transport. The 'iron triangle' of vested interests consisting of civil engineering companies, the politicians of the conservative parties and the bureaucrats has been working effectively to promote this process. Lately, this collusion has taken a new policy guise of creating 'national land axes', consisting of a bundle of trunk Shinkansen and expressways to form spinal corridors of the nationwide transport network. Some Japanese geographers, including T. Yada (1999), the president of the Japan Association of Economic Geographers, have been providing academic glitter to the 'national land axis' development project.

Figure 9.3 A promotional leaflet for 'torenta-kun' prepared by JR – a convenient connection between the Shinkansen service and rent-a-car is emphasized, to the neglect of rural conventional rail service. (The Japanese text in the balloon reads: 'Drivers and all the passengers of our rent-a-car enjoy a discount on JR fares and charges – Fare 20% off, Express charge 10% off, Luxury Car Surcharge 10% off.)

The trend of passenger-kilometres and foreign investor influence on the neo-liberal corporate strategy of the privatized JR – the case of JR West

In December 2001, the government lifted all the restrictions that had stood in the way of full privatization of the three JR passenger companies on Honshu Island. The JRs on Honshu Island, now much closer to pure private companies, began to pursue their profits even more aggressively. An example is JR West, whose corporate strategy has recently been largely influenced by foreign investors who own as much as 11 per cent of its equity. The operation of JR West can be divided into three categories: Shinkansen between Osaka and Fukuoka (Hakata), the Kyoto–Osaka–Kobe metropolitan sectors where JR West faces fierce competition with parallel private railway lines, and other rural sectors with sparser population. The total passenger-kilometres travelled on JR West increased by almost a quarter from the year of privatization to 1993 (Figure 9.4).

The recent stagnation of the Japanese economy must in part be responsible for the general decline of passenger-kilometres thereafter. Yet regional

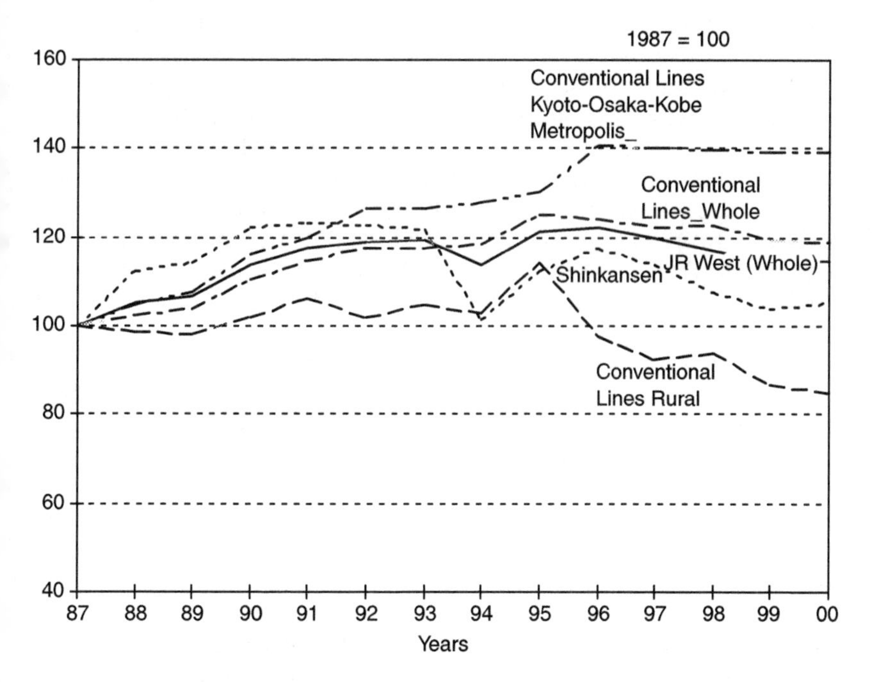

Figure 9.4 Changes in passenger-kilometres of JR West by various types of lines and geographical areas

Source: Unpublished data of JR West.

disparity is remarkable. The increase by 40 per cent of passenger-kilometres travelling on conventional lines in the Kyoto–Osaka–Kobe metropolitan area appears in stark contrast to the stagnation and recent decline (down 17.5 per cent over six years from 1994) of passenger-kilometres travelled on conventional lines outside the metropolitan area. This suggests that JR West succeeded in grabbing passengers from parallel rival lines through aggressive corporate strategy in the Kyoto–Osaka–Kobe metropolitan area. On the other hand, the stagnation and recent decline on conventional lines outside the metropolitan areas indicates that the competitive edge of railways against other modes of transportation has been disappearing. It is interesting to note that, although Shinkansen passenger-kilometres showed an increase of 23.3 per cent over the four years up to 1991, it has suffered decline ever since, indicating its weaker competitive position against airlines (Kasai, 2001).

The foreign shareholders have been demanding that the company disinvest from rural sectors where the operational cost is several times more than the revenue ('JR West is Rationalising 13 Rural Lines' *Asahi Shimbun*, 21 December 2001). To meet their demand, JR West has been adopting a shrinking management policy in operating the deficit-generating rural sectors, which have suffered 11 per cent reduction in number of passengers over the last eight years. The actual measures include further reduction in frequencies of passenger services, increasing number of trains without conductors and unmanned stations, reduction of the maximum speed of trains on the rural sectors, total suspension of passenger service during daytime on a particular day of the weekend on the pretext of 'track maintenance', and so forth. These management policies are now being applied to a wider geographical area, to cover eventually 30 sectors, or 30 per cent of the total JR West network in terms of the rail distance. JR West claims that this strategy will cut the cost of operations by 30 per cent, while it will make the rural services more dysfunctional, decisively depriving railways of their competitive edge against automobiles.

For foreign investors who regard proprietorship of the railways merely as fictitious capital, the use value of railways is at best of secondary importance, and at worst a negative factor that erodes profits. The social embeddedness and the historical heritage that the railways have had in the localities for decades are totally outside the consciousness of foreign investors. The motive to procure abstract money out of railway capital thereby imposes a specific destiny on every locality, and often generates a deep sense of deprivation among the local populace.

Concluding remarks: why railways?: the rationale for revival of a nationwide homogeneous rail network

Privatization of the Japanese railways categorically did not succeed in bringing people to patronise railways more. Rather, the introduction of

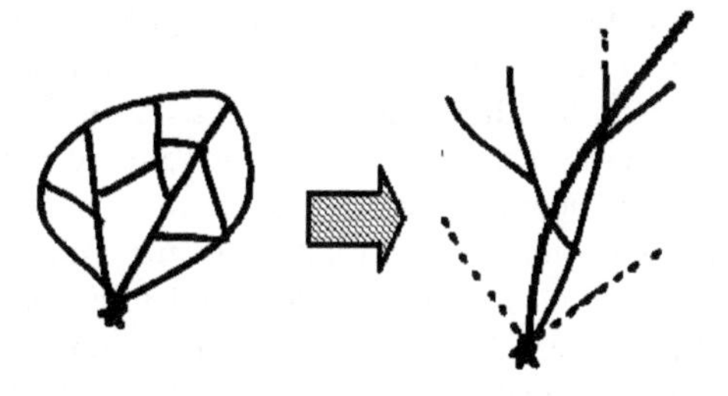

Figure 9.5 The changing spatial configuration of the rail network in Japan

laissez-faire market mechanisms has led operators to a more aggressive 'cream skimming' strategy in the urban areas and concomitant decline of services due to disinvestment in the rural areas. The strategy put into practice to strengthen corporate finance has caused a spatially more heterogeneous railway service and a network with stronger nodality very different from those under the management of JNR (Figure 9.5).

Swift as the journey is along the 'national land axes' between large metropolitan areas, connectivity between different points in rural areas has greatly deteriorated, generating increased spatial disparity in Japan, and resulting in lengthened journey times by rail between rural towns.

This is the reality behind the myth of 'success' in privatizing the Japanese railway system. Here, some may ask, 'Why still insist on railways?' One might as well travel in his or her own private or rental car'. Indeed, there are several rationales why we should retain railways even in rural areas. If the railway is gone, the hardest-hit will naturally be those who rely on public transportation almost exclusively: the elderly, the poor, the school children and physically handicapped people who cannot own or drive a private car. Without public transportation organised to form a spatially homogeneous network, the extent of their action space will be severely limited, and many public facilities will be beyond their spatial range.

The railway is more environment-friendly and emits less CO_2, believed to cause global warming, than private cars. The CO_2 emission of a diesel rail car is 557 g-C/km (g-C/km is the unit that indicates carbon emission in grams per kilometre of carriage operation) for JR Hokkaido, 476 g-C/km for JR East and 637 g-C/km for JR West. These compare favourably to emissions from a single passenger car at 75 g-C/km (see Kamioka, 2002). This means that if a diesel railcar transports just a little more than 10 passengers per car, a diesel rail car is more environment-friendly than a passenger car. The figures become more favourable to rail if electric rail cars are deployed. Unless a local branch line is disparately isolated, rail is almost always the wiser choice to protect our planet.

The 11 September (2001) incident in the US manifested the physical vulnerability of the passenger airlines and Amtrak immediately demonstrated its raison d'être in the United States. A newspaper editorial appeared in Philadelphia stating as follows: 'Amtrak carried medical supplies, emergency personnel and victims' families to New York City after the tragedy. It hauled tons of extra mail across the nation. And the railroad helped airline ticket holders by scheduling scores of additional trains'. There seems little doubt that fear of flying will increase Amtrak's ridership and revenue in coming years'. ('Attack makes clear Amtrak has a role' *Philadelphia Enquirer*, as quoted in *UTU Daily News Digest*, 24 September 2001).

Ironically, however, privatization of railways has been eroding their competitive advantage in terms of safety. The case of a passenger having fallen off the platform in a station in downtown Tokyo was the worst case in point and widely reported in Japan. Korean and Japanese passengers attempted to rescue the passenger who fell on to the track, but they were themselves run over and killed by an approaching train. Overcrowded Japanese commuter services allow men to 'give women the feel' on a train, or women to blackmail men on false charges of having been given the feel. These situations have generated worries about security among rail commuters.

'Economic efficiency' is a criterion that should not be understood without taking all of the above factors into account. The performance of railways should not be measured using an indicator of short-run economic performance only. The railway has the best potential to materialize efficiency in a real sense if it is managed properly with the users participating in democratic decisionmaking processes.

Note

1. Those books of which titles are shown in Japanese are written in the Japanese language. English translations in parentheses are provided by the authors of this chapter for the convenience of international readers.

10

Developing Public Transport in Indian Cities: Towards a Sustainable Future

Swapna Banerjee-Guha

Introduction

Since the Habitat II Conference of the United Nations Conference on Human Settlements (UNCHS) in Istambul in 1996, the concepts of 'sustainable cities' and 'sustainable development' have gained ground. While admitting that 'sustainability' needs to be a key word in the formulation of development initiatives, it is equally important to understand the meaning of the term in the context of the countries of the South. In the macro-context of globalization and liberalization of recent decades, sustainability in development initiatives is facing a grave crisis all over the world, especially in such countries.

Concepts of 'sustainable development' and 'sustainable cities' for the countries of the South essentially need to address much wider parameters in order to prioritize the issue of poverty and redistribution. Sustainability in these countries must also consider the underlying economic, social and political attributes that often tend towards exclusion of the economically weaker sections of society who suffer the most from degraded and polluted environments (Low and Banerjee-Guha, 2001). The exclusionary tendency of the 'sustainable city' concept of the World Bank and other multinational agencies, with a thrust towards techno-managerialism, is actually devoid of the multidimensionality of development and the synergistic relationship obtaining among different policy sectors such as employment generation, poverty reduction and resource recycling. It is time for city sustainability in the South to be focused upon a socially and politically just philosophy having a balance among environmental sustainability, social equity and economic growth.

The United Nations Conference on Human Settlements in 1976 had recommended transportation policies for developing countries with a priority for public transport to achieve maximum benefit for the maximum number of people. The focus was on achieving efficiency with minimum transport-related degradation of the environment and optimum protection of non-renewable resources (Patankar, 2000: 111). The United Nations Conference on Environment and Development (UNCED) in 1992 in Rio de Janeiro reaffirmed the same priorities by integrating the issues of sustainable human settlements with transport programmes favouring high occupancy public transport. Nevertheless, in many developing countries including India, effective policies for urban public transport were not properly formulated. Rather, in the post-liberalization era there arose a widening gap between demand and supply in the sphere of affordable urban mass transit systems, with a proliferation of personalized modes of motor transport. At present, the 23 1,000,000 plus (population) cities of India, with a minimum of 8 per cent of the national population, account for 33 per cent of the total vehicles of the country.

The paradox of urban transport planning in contemporary India needs to be understood in the above context. While programmes for developing, overhauling and improving public mass transit systems abound, the liberalization lobby within the government vigorously promotes road-based privatized transport projects, disregarding the societal and environmental consequences. A 'road rage' lobby enjoying strong governmental support has emerged in present day India pushing mega-transport projects in the large cities.[1] The focus of 'sustainable cities' and related programmes thus gets shifted.

Further, in India the concept of 'sustainable city' itself is also not free from confusion. During the early eighties in the days preceding the New Economic Policy, India opted for 'Sustainable City Programmes' (SCP). Originally formulated by the United Nations under its Environment Programme, they were drafted as 'Environmental Guidelines for Settlement Planning and Management' (EPM). Replicating such programmes of the North they had a focus on self-reliance, the market mechanism, city redesigning, increasing energy efficiency and liveability (Haughton, 1997: 192). In fact a narrow environmental bias (Satterthwaite, 1996: 33) grew in India inevitably deprioritizing the basic needs of the majority of the population.

The point that has been systematically overlooked by a large group of Indian transport experts is the detrimental role of development models transferred from the North (as exemplified in the policies of the European Commission: see Chapter 7 of this volume). Increasing problems of sustainability in development projects have actually been found to be closely related to such models which are vigorously advocated by the international multilateral agencies for various countries of the South. One clearly finds the above trend accentuated in the recent Structural Adjustment Programme (SAP), with private investment indiscriminately encouraged in the sphere of basic urban infrastructure including transport. A heavy bias towards commercialization is becoming associated with this key sector, leading to increasing numbers of personal vehicles – of which congestion and pollution are the inevitable outcomes. A new trend of environmental activism launched by affluent citizens' groups in mega-cities of India has come to occupy centre-stage. This trend is nothing but an off-shoot of the techno-managerial approach to sustainable development associated with elitism. It systematically overlooks the vital questions of *social* sustainability of the urban environment taking the entire cross-section of society into consideration. The planning and organization of urban transport in India is replete with such contradictions.

Urban transport in a developing country is a complex phenomenon. It is not possible to cover all its aspects in this one chapter. In this chapter the intention is to analyse the status of the transport system in urban India with regard primarily to road transport and with special reference to the issue of sustainability and people's needs.

Problematics of urban transport in India: an overview

Indian transport policies during the five year plans were too general, with scant attention to the actual problems. It was only after the Sixth Plan that urban transport was considered as a separate item. From the Seventh Plan onwards to the introduction of the new economic policy (NEP), the main focus was given to public transport. Since the 1990s the urban transport sector in India, like many other infrastructure sectors, has experienced an excessive thrust of private investment and private modes of motorized transport. On the other hand, because of the lack of self-finance and the required support from the State, mass transit services have for a long time not been keeping pace with the rising demand of the people. The inability of the urban transport system to address the broader issues of socio-economic development is already a grim reality. Amidst this, a two pronged attack on mass transit has been launched. While a motorized transport agenda is pushed forth with considerable State support, a distorted concept of sustainability, overlooking people-centred approaches and welfare concerns for the majority is simultaneously popularized. As a result, since the 1990s, the major contributor to the haze and poison in the air was no longer the factories but automobiles (Sharma, 2000: 107). In 2000 two-wheelers and cars respectively contributed 78 per cent and 11 per cent of the total vehicular pollution in the large cities of India. Increasing congestion has also been an added problem.

How, then, would one proceed to develop an understanding of the urban transport scenario in India? To start with, one needs to realize that transport and land-use patterns in Indian cities, as in any city of a developing country, have distinctive but closely interrelated characteristics. High densities, intensely mixed land-uses, short trip distances, a large share of walking and a smaller share of non-motorized transport are the interrelated features of most Indian cities (Tiwari, 1998: 143). Compared to their Western counterparts, they obviously consume less transport energy and can be termed 'low-cost strategy' cities. Further, the higher densities and mixed land-use patterns are especially conducive to public transport trips and shorter trip lengths, leading to a reduction in the number of motorized trips. While in western cities these features are only achieved by comprehensive planning, zoning and a selective tax system, in Indian cities these have evolved regardless of such efforts. On the other hand, due to intense class variations, different road users in such cities have different but conflicting requirements. While pedestrians and cyclists need safe road space and shaded pavements, private vehicular traffic prefers uninterrupted flow with minimum delay at intersections. Public transport requires frequent stopping facilities. The road infrastructure, however, does not provide the basis for such complex and conflicting demands, forcing all types of users to accept less than optimal conditions. The markedly evident spectre of poverty associated with imbalance in the distribution of urban services and differential access to resources all have their repercussions on transport. These cities therefore warrant a perspective other than that used in the analysis of cities in highly motorized countries (HMCs).

A large number of non-motorized and motorized vehicles along with trains characterize the transport system of Indian cities, between 45 and 80 per cent of the registered vehicles are two-wheelers, while cars account for between 5 and 20 per cent (Table 10.1).

Table 10.1 Vehicle composition in selected Indian cities, 1994

Cities	**Vehicles ('000)**	**Vehicle fleet composition (%)**						
		MTW	**Car/ Jeep**	**Taxi**	**3 WH**	**Public bus**	**Truck**	**other**
Bombay	689	38	42	6	4	1	6	3
Calcutta	536	40	40	4	1	3	8	5
Delhi	2543	69	20	1	4	1	6	0
Chennai	937	70	22	1	1	5	3.5	2
Hyderabad	543	81	9	0.4	3.6	1	4	1
Bangalore	716	75	16	0.5	2.6	1	3	2
Ahmedabad	477	72	11	0.2	9.8	2	2	1
Pune	300	75	9	1	6	1	7	2
Kanpur	208	85	8	0.03	97	4	3.6	1
Lucknow	266	78	11	0.3	1.7	1	3	5
Jaipur	339	66	12	1	2	4	6	9

Note: MTW: Motorized two-wheelers; 3 WH: Three-wheeled scooter.

Source: Motor Transport Statistics of India, Ministry of Surface Transport, Government of India, 1995.

The road network is used by almost seven categories of motorized and non-motorized vehicles. In mega-cities public transport constitutes the predominant mode of motorized transport with buses carrying 20–65 per cent of the total trips, excluding trips by walking. Besides buses, suburban trains account for a major share of movement in cities like Mumbai and Kolkata. In Mumbai public transport modes carry 83 per cent of the total passengers during peak hours (D'Monte, 2001: 4). The public preference for using public transport is quite high, and in cities with comparatively better public transport, such as Mumbai, Chennai, Bangalore and Kolkata, the number of users is extremely high, including those whose income is 50 per cent more than the average income in the cities. However, despite the above significant share of work trips, public transport is not currently sufficient in any of these cities, with an overwhelming gap between demand and supply.

Notwithstanding this gap, a clear thrust towards policy oriented to personalized vehicle transport is apparent, with an associated increase in the number of such vehicles in the recent decade. In Hyderabad, for example, in 2000 one million personalized vehicles were operating for five million people, of which 70 per cent were two-wheelers. Vehicular density in the metropolis was an astounding 220 vehicles per road kilometre (Venkateswaralu, 2000: 139), The current automobile explosion has drastically changed the quality of air in large cities of India. Research on various cities have reported that more than 50 per cent of the pollution in cities is accounted for by vehicles. The link between traffic organization and the growing pollution hazard has not been considered seriously by the planners. Had it been, a greater thrust would have been given to public transport, both rail and road, with clean fuel technology and fewer vehicles on the road.

Instead, the last decade has seen a tremendous rise of vehicular traffic and resultant pollution in the major cities of Mumbai, Delhi, Bangalore, Chennai and Hyderabad. According to a recent study of Rail India Technical and Economic Services (RITES), two-wheelers and cars in Indian metropolises now respectively contribute 78 per cent and 11 per cent of vehicular pollution. Three-wheelers contribute around five per cent, buses two per cent and trucks four per cent (Patankar, 2000: 112). The growth of motorized two and three-wheelers has undoubtedly been a prime source of congestion, pollution and high fuel consumption in almost all Indian cities. With public transport being inadequate and ownership of cars still remaining beyond the reach of the majority, two-wheelers appear a popular option with nearly 1.8 million two-wheelers being produced annually. Almost 60 per cent of the personal vehicles in the large cities are in this category. Of the total of vehicles in India in 1997 cars and jeeps constituted 12 per cent while two-wheelers constituted more than 60 per cent (Reddy, 2000: 120). In subsequent years, their number has increased at a much higher rate than public transport vehicles and their impact is being systematically felt on the health and transport situation in both mega and second order cities.

Urban transport planning in India thus needs to be an integral part of urban development and environment planning. The primary issues can be identified as, (i) sustainability of urban development, (ii) sustainability of urban transport, (iii) social equity in public transport, (iv) environmental degradation due to transport, (v) financial sustainability of public transport, and (vi) institutional aspects of public transport. The following sections offer a brief picture of the transport situation in selected cities of India.

Public transport and related problems in selected Indian cities

In most Indian cities 25 to 30 per cent of the total volume of composite trips involves walking. The situation varies from city to city but in most cases facilities available to pedestrians are very poor and public transport systems are mostly inadequate. Let us consider briefly a selection of cities.

Delhi the capital city of India

Delhi has an extensive road network with a total length of 20,487 kilometres of which 880 kilometres are more than than 30 metres wide. The *per capita* availability of roads in Delhi is higher than many cities of Asia and Europe. However, cities in the United States and Australia have still higher *per capita* road length but simultaneously high levels of motorization and inadequate public transport systems. On the other hand, although Delhi has one of the lowest rates of vehicles per kilometre; 65 per cent of its vehicular fleet consists of motorized two-wheelers which fill up the available road space in a somewhat chaotic manner causing considerable congestion. Railway services, including the Ring Railway EMU, have an extremely limited capacity.[2] Public transport is comprised mainly of the buses of the Delhi Transport Corporation and private buses. However, the inadequacy of these public transport services is well known, and auto-rickshaws offer a parallel service as an intermediate public transport system.

Mumbai, capital of Maharashtra

Mumbai is the commercial capital of India. In contrast to all other Indian metropolitan cities, the suburban (surface) railway is the most developed public transport facility, carrying 40 per cent of trips by motorized modes. In the metropolis and its sister city New Bombay the bus service (accounting for another 40 per cent of total trips by all vehicular modes) is provided by the Bombay Electric Supply and Transport Corporation. State transport buses also offer a limited service in the city.

Kolkata, capital of West Bengal

Public transport is provided by buses, trams and suburban trains. Buses and trams are run by Kolkata State Transport Corporation (or by the private sector) and Kolkata Tramways Corporation respectively. However the surface trains, unlike Mumbai, only partially cater for the city areas. In Mumbai, the suburban service caters for the entire north–south alignment through western, central and harbour lines whereas in Kolkata eastern and south-eastern lines do not enter the city of Kolkata while the north-eastern and southern lines cater to a small area in the eastern and southern parts of the city. The biggest carrier is the Eastern Railway suburban service which terminates at Howrah and does not enter Kolkata. Since 1985, the underground suburban metro service has been introduced and is the first of its kind in India. Running on a stretch of 16.45 kilometres, the contribution of this mass transit system to the transport situation has been substantial and caters for about 25 per cent of the public transport users of Kolkata proper.

Chennai, capital of Tamil Nadu

Besides the scanty suburban trains, a bus service, provided by the Pallavan Transport Corporation, provides the major structural base of public transport, accounting for 63 per cent of the total trips by all vehicular modes.

Hyderabad, capital of Andhra Pradesh

The train and bus services combine to form the public transport system. Neither of these services is as expansive as they are in Mumbai or Kolkata. The population in all these cities, including the largest two, largely depends on other modes, such as, auto-rickshaws, etc.

Bangalore, capital of Karnataka

The transport pattern is radial, following the layout of the metropolis. The railways do not provide any effective suburban coverage, creating an excessive load on the bus service provided by Karnataka State Road Transport Corporation (KSRTC) and Bangalore Transport Corporation (BTC). The service caters for almost 70 per cent of the total person trips by vehicular modes (TCS, 1993: 91) and still proves inadequate, leading to a proliferation of personalized vehicles.

In all these cities, and many second order metropolitan cities such as Kanpur and Lucknow in Uttar Pradesh and Nagpur and Pune in Maharashtra, there is considerable dependence on alternative modes like the auto-rickshaw and personalized modes, especially two-wheelers (Sharma, 1985: 62). The use of the mass transport system in second order metropolises such as Lucknow is extremely low (3.61 per cent) compared to the larger metropolizes of Mumbai (80 per cent), Chennai (67 per cent) and Bangalore (70 per cent).

Major problems related to transport in the large cities of India are:

1. the inadequacy of public transport,
2. the disproportionate relationship between road length, people and the number of vehicles (see Table 10.2), and
3. increasing pollution hazards.

Table 10.2 Vehicular growth in selected cities of india

Category	Motorized vehicle		Two-wheelers		Cars, jeeps & station wagons		Freight vehicles		Non-motorized vehicle	
City	No. of vehicles	Growth (decadal)	No. of vehicles	Growth (decadal)	No. of vehicles	Growth (decadal)	No. of vehicles	Growth (decadal)	Vehicle type	Year
Mumbai	*1981* 308,881 *1990* 609,904	97%	*1981* 78,474 (25%) *1990* 231,932 (38%)	196%	*1981* 150,711 (49%) *1990* 258,315 (42%)	71%	*1981* 38,447 (12%) *1990* 48,556 (8%)	26%	 Bullock carts Handcarts Bullock carts Handcarts	*1981* 802 2,941 *1990* 418 16,791
Madras	*1981* 103,666 *1990* 409,632	295%	*1981* 58,435 (56%) *1990* 280,800 (69%)	381%	*1981* 30,646 (30%) *1990* 90,409 (22%)	195%	*1981* 5,362 (5%) *1990* 18,935 (5%)	253%	Bicycles Bicycles ownership/ household Handcarts Bullock carts	*1981* 152,000 (0.27%) *1990* 535,000 (0.61%) 1978–1984 Decrease of 51%
Bangalore	*1976–77* 108,437	204%	*1976–77* 62,199 (57%)	281%	*1976–77* 23,808 (22%)	152%	*1976–77* 6,081 (6%)	87%	Bicycles (% total person trips)	*1966* 32.3% *1977* 18.2%

	1986–87 329,255		*1986–87* 236,726 (72%)		*1986–87* 60,007 (18%)		*1986–87* 11,336 (3%)			*1982* 16.1% *1988* 4.2%
Lucknow	*1981–82* 58,230	271%	*1981–82* 42,724 (73%)	305%	*1981–82* 8,410 (14%)	172%	*1981–82* 2,786 (5%)	51%	Bicycles, cycle-rickshaw and others	*1980* 80 *1982* 76
	1990–91 215,853		*1990–91* 173,186 (80%)		*1990–91* 22,917 (11%)		*1990–91* 4,209 (2%)			*1984* 79 *1990* 68

Notes
Motorized vehicles include two-wheelers, cars, taxis, buses, trucks, tractors and tailors.
Figures in brackets indicate percentage of total.

Associated economic, social and operational aspects of the transport problems also need to be considered. The related economic aspects are (i) increasing cost of production and prices of goods due to inadequate provision of transport infrastructure and services, (ii) decreasing productivity of human resources due to low availability and quality of transport, (iii) low income status of people affecting the economy of public transport and transport costs, (iv) resultant lack of expansion and improvement of public transport, (v) lack of revenues affecting maintenance and (vi) lack of funds for investment in public transport. Social problems include (i) lack of equity in the distribution of transport costs and benefits, and (ii) lack of importance given to the transport needs of the urban poor. The operational problems include (i) marginal increase in road space for public transport, (ii) lack of road maintenance, and (iii) low productivity of transport operations (TCS, 1993: 79). All these problems are interrelated and have become accentuated in the post-liberalization years, with the increasing policy bias towards personalized transport and related mega-infrastructural development, and its resultant increase in pollution and associated health hazards.

As Mumbai epitomizes the contemporary aspirations of Indian cities to enter the global urban system, a critical analysis of the priority areas of urban transport planning in Mumbai may throw light on the philosophy and methodology of contemporary urban transport planning in India.

Contemporary transport planning in Mumbai

Contemporary transport planning and projects in Mumbai essentially reflect basic contradictions between need-based planning and economic globalization priorities. For example, while programmes are being undertaken for developing and improving the mass transit system at governmental levels in the pubic sector, the 'liberalization' lobby of the same governments promotes road-based, privatised vehicular transport projects in complete disregard of the environmental effects. The powerful road lobby of Mumbai, a feeder group of the national lobby, enjoying considerable State support under the New Economic Policy, has been instrumental in pushing forward mega-projects that aggravate environmental degradation of the city and cause deterioration of the health of its residents.

Mumbai is an elongated city having a north–south expanse. Because of this physiographic configuration, linked with its root in seven linear islands reclaimed and joined together, Mumbai's entire transportation network reflects an overemphasised north–south alignment without sufficient east–west links. This has given rise to undue detour of rail traffic, but even more of road traffic, resulting in longer hours of travel, overuse of fuel and environmental pollution. Since the initial years of national planning, the transport network of Mumbai has been the responsibility of the public sector. The number of organizations and agencies sharing the responsibility for

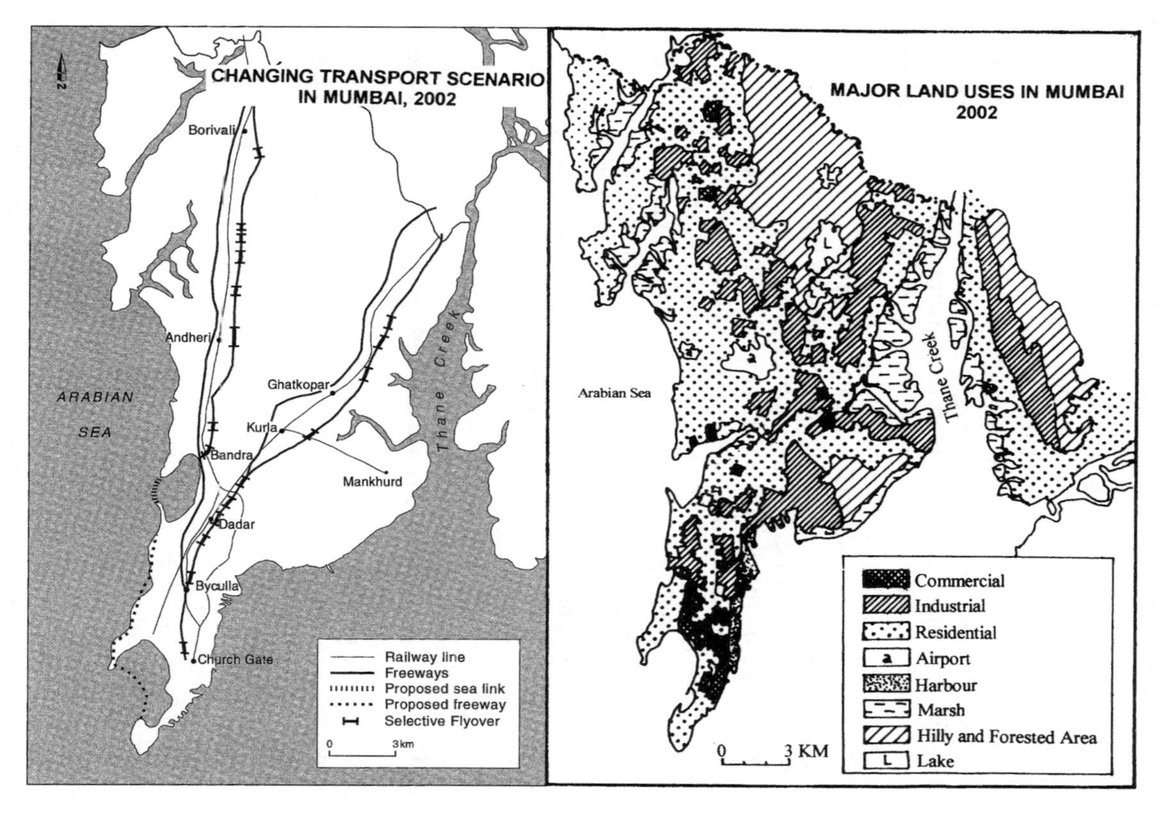

Figure 10.1 Maps of Mumbai showing the transport scenario 2002 and major land uses

construction, design, operation and maintenance of the transportation services are many, almost 13 in number, having among them a fragile co-ordinating network. Some of them like the Maharashtra State Road Development Corporation (MSRDC) is fully owned by the Government of Maharashtra even though it has been incorporated as a limited company since 1996 under the Indian Companies Act.

The Greater Bombay Development Plan of 1973 (BMRDA, 1973: 71) envisaged a multi-nodal structure for the Bombay Metropolitan Region having New Bombay and Kalyan as major regional growth centres. The road and suburban rail networks had, up to then, followed the alignment of the colonial transport infrastructure. The extended rail network has five rail corridors, two on the Western and three on the Central Railway, running about 2000 trains per day. Similarly, the road network has three main north–south corridors, the Western, Central and Eastern, the latter carrying the heaviest truck traffic. Actually the road network of Mumbai is organic in nature, constructed before the advent of automobiles. The north–south alignment has been dictated by the lie of the land but it serves the entire traffic volume heading towards the central business district located in the extreme south. In the mid-eighties, with the launch of the Maruti – 'the common man's car' – vehicle ownership increased in all cities, especially in Mumbai. The road lobby for the first time gained ground, helped by easy finance for vehicle purchase provided by non-banking financial companies to make personal ownership of motor transport easier. The resultant expansion of the domestic car market led to a steep rise in the number of vehicles in Mumbai, bringing pollution and environmental deterioration in its wake.

Problems of transportation in Mumbai are also due to its faulty urban form. The initial concentration of commercial, finance and office sectors in the southern tip of the metropolis was a colonial design which was never critically questioned although attempts were made to reconstruct it later with a multi-nuclear pattern. Rather the subsequent nexus of State, industrial and commercial capital, supported by the builders' lobby, persistently worked towards intensifying the use of land in Southern Bombay aggravating the pressure on transport and other related infrastructure.

The two key comprehensive studies with wide ranging proposals on transport development in Mumbai were the 1981 Wilbur Smith study and the 1994 W.S. Atkins study. In between, various central and State government departments and private consultants' groups in several studies have written about the need for east–west road links and free passage ways at vital intersections of the express highways in the north-central and north-western parts. Interestingly none of the studies advocated any major road improvements in the southern city because of high cost. In all these reports the needs of pedestrians were also by and large ignored while demand for personal vehicular traffic was specially highlighted. The first Bombay Urban Transport Project (BUTP), undertaken in 1977 and completed in 1984 at a cost of Rs 390 million with a World Bank loan of US$25 million, however, focused on the need to improve the bus network. This led to the construction of a number of road over-bridges. The second phase of the BUTP, started in 1985, also had the objective of enhancing the capacity, efficiency and financial viability of the urban public transport system in Mumbai metropolitan region, particularly the mass transit system, with suitable policies and appropriate investment in transport infrastructure. Following this, in the early nineties, rail improvement initiatives were geared up in the form of improvement of the existing railway system and simultaneous expansion to newer areas of the city. Even though a road link had already been constructed between Mumbai and New Bombay in 1973 through a fragile road system and an inadequate road bridge near Vashi (the major node of New Bombay), the much awaited appropriate railway link and a proper road bridge could only be established during this period. Simultaneously extensive rail improvement works were undertaken on the Western, Central and Harbour lines at different locations throughout the city. The contradictions of such programmes, however, surfaced in the large scale displacement and marginalization of innumerable poor families from near the railway track, whose rehabilitation was a reluctant responsibility of all the concerned organizations.

The contradictions of globalization and liberalization subsequently surfaced. During the late 1990s a drastic policy shift was made by the liberal government through the initiation of a range of road schemes opting for private, motorized transport in preference to public transport. The hidden agenda of the Mumbai Urban Transport Project (MUTP II, 1998) showed up. In the late 1990s, falling in line with the dictum of the New Economic Policy, the State Government, bypassing the apex planning body of the Mumbai Metropolitan Regional Development Authority (MMRDA), entrusted both the planning and execution of the road projects to the Maharashtra State Road Development Corporation (MSRDC) which had no expertise to take on the work. The total cost of all these schemes was over Rs80 million. Even W.S. Atkins, the consultants hired by MMRDA to advise on a comprehensive transport plan for Mumbai metropolitan region in 1994, argued that Mumbai was unique in its emphasis on public transport which carried 83 per cent of all passengers during peak hours. Private cars and two wheelers accounted for 8 per cent and 9 per cent respectively. Ironically, in the contemporary transport programmes of Mumbai, the latter group has received all the attention (D'Monte, 2001: 3).

The three major pillars of the contemporary transport programme in Mumbai are (1) road flyovers (2) sea links and (3) freeways, all encouraging personal traffic. Even Konkan Railway that has been a major link between Mumbai and its expanded rimland in the south-western part of Maharashtra is contemplating a sky-link for speedy traffic for elite passenger groups. Following recommendations by the American consulting firm Wilbur Smith, 50 road flyovers are being constructed within the city of Mumbai. Space underneath some flyovers is even going for commercial use, belying the prescribed benefits of relieving congestion on the road. While the State Government systematically projects flyovers as the ultimate solution to the problem of congestion and vehicular pollution, vehicles plying at a higher level and speed aggravate pollution of the adjoining residential buildings and create conditions for more cars to come. Cars use more road space, leaving a much smaller road area for public transport such as buses which are forced to resort to reduced speed. The latter cannot use the flyovers as these are being constructed at major road intersections where bus stops are provided on the main road. Even the World Bank, the principal funding agency of MUTP I and II, has opposed the flyovers for their negative impact on the environment (noise and air pollution), on public transport and affordability of transport for the general public.

Several committees (TCS, 1998: 16) have negatively opined against the flyovers and citizens' groups have fought through the courts against the institution of tolls. Ultimately the legality of these tolls has been upheld by the Mumbai High Court. Following this decision, the State Government has started charging tolls on only the entry point flyovers to the city which has given rise to many other critical issues as the entire financial burden of the flyovers funded from public revenue has finally been equated with the benefit of a meagre 17 per cent of the population (D'Monte, 2001: 4).

The second pillar, the proposed 4 kilometres long Worli Bandra sea link aiming to save time for vehicular traffic, has invited serious criticisms from environmental groups. With more than 100 acres of land already reclaimed and a few hundred waiting for reclamation, one wonders why it is promoted as a sea link. The prospective displacement of the fishing community in the adjoining Mahim Bay and the endangering of the ecosystem due to the possible loss of mangroves are associated sore points. As people are not allowed access to the plan documents of the present link and the future extensions, the Indian People's Tribunal (a citizens' collective initiative) decided to hold a public hearing on the issue and submitted a report condemning the project. All the erstwhile committees appointed to scrutinize the merit of the link had also opposed it, indicating that it would add to serious traffic congestion in a large area of South Mumbai. The People's Tribunal identified the project as a violater of environmental laws, capable of giving rise also to negative social impact. The project could prove extremely detrimental to the environment causing erosion at several areas and harmful effects on the mangroves. The report stated that no proper study of environmental impact was done, local fishermen who feared displacement were not consulted and feasibility studies proved outdated.

The most controversial road project in Mumbai is the third item, the freeway project, again recommended by Wilbur Smith four decades ago (D'Monte, 2001: 4). It is interesting to note that recommendations that were shelved then are now being revived with considerable fanfare as modern urban development programmes. A 15 kilometre road which envisages a bridge from Worli across the Hali Ali Bay, a coastal freeway up to Malabar Hill and a third bridge across the bay at Marine Drive to Nariman Point, all located in Southern Mumbai, are the major features of the project. By contributing to further congestion and slowing down car speed in South Mumbai (the car speed already has already gone down from 20 kph in 1993 to 12 kph in 1997), encouraging more car traffic into the area, one wonders how this project can be expected to help the majority of the public.

A recent City Transport Scheme for Mumbai, jointly formulated by Central Government, Maharashtra State Government, Bombay Electric Supply and Transport (BEST) and the City Police, has suggested reasonable and sensible measures for improving suburban rail services including the introduction of air conditioned coaches to cater for the affluent class. Road widening, footpath construction and maintenance, underground crossings, subways, overbridges and improvement of the road traffic signalling system are their additional suggestions. Finally, provision of adequate finance, according to these authorities, is the basis for making all the programmes successful. The major thrust of such policies is always the strengthening of suburban railways and public road transport (Low and Banerjee-Guha, 2001).

It is, however, clear from the nature of ongoing transport plans of Mumbai that liberalization options have considerably influenced the development of urban transport in this and many other mega-cities in India. The negative impacts of such transport policies are patent and numerous in various countries, yet these cities plan to go through the same pitfalls and formulate transport plans thoroughly alienated from people. Essentially three basic questions surface from the contemporary transport planning in India. First, what should be the indispensable level of urban transport provision to meet the welfare objectives of the majority and, therefore, what kind of transport infrastructure would be needed within the limits of resource use? Secondly, how should the urban transport sector be managed to be self-sustainable without undermining the sustainability of other sectors of the economy? Finally, how would the organization of human activities add to the sustainability of a city's transportation, its impact on human health and its demands on non-renewable resources (Patankar, 2000, p. 113)? All these questions are vital to address in order to make the urban transport planning agenda strongly committed to the concept of sustainability.

Towards a sustainable future

Privatization of urban transport, especially of roads, figures as a major item of the contemporary transport scenario in India. This idea, however, is not very recent. Since 1985, the Government has been floating the option under the popularly known BOT (build, operate and transfer) approach. With increasing constraints on public sector finance, this was considered a panacea for infrastructure investment shifting to the private sector the responsibility for developing the infrastructure through open market borrowings, following which the facility user price would be decided on commercial principles. Accordingly, projects that became popular were (a) expressways, (b) mega-bridges and tunnels, (c) widening of existing roads to four lanes, and (d) large road flyovers in cities.

Adherence to policies supporting commercialisation of investment in road transport would mean that no road, by-pass or bridge would be available for free passage either for passenger or for freight traffic. On the one hand, this would certainly put heavy pressure on the public, while on the other the entire exercise would prove economically unfeasible unless the corridors have very high traffic densities. The decision to invite real estate development activities to mitigate the prospective loss (if the corridors do not generate enough traffic) also became a critical issue. It is anticipated that unless the government takes a proactive role in the regulation and provision of land to real estate developers, proper returns on investment from such mega-transport schemes would remain unrealized. At present the 30 road projects identified under BOT schemes are facing complications due to constraints of insufficient finance, lack of private interest as well as absence of prioritization agendas on the part of the government. Even an Asian Development Bank sponsored study (ADB, 1991: 26) in the early nineties suggested that priority should be given to improve and upgrade the existing road network throughout the country, especially the inter-city linkages. Despite these suggestions, the budgetary allocation has been minimal leading to increasing loan assistance from international agencies (Patankar, 1994: 58). A more reasonable option suggested by transport experts is fair competition between the public and private sectors in the provision and maintenance of intra and inter-urban transport, achieving optimum benefits without monopolistic control (Sriraman, 1997: 20).

The significance of public transport needs to be reiterated. The present policy for urban public transport in India is extremely limited in its approach, coverage and viability. It is also not capable of self-financing its own development to meet the evergrowing demand. Transport policies thus need to adopt an inter-sectoral approach which is sufficiently comprehensive, demanding a strong management system and inter-agency co-operation. Patankar (2000: 110) suggests a philosophy of Integrated Human Settlement Management for providing the necessary framework for urban transport policy planning. Following this, the supply and demand sides of the transport equilibrium need to be developed on the basis of regulated, resource-efficient and affordable transport modes. The primary content of the balanced and integrated transport system that it would lead to can only be realized by a viable public transportation system which would automatically limit growth of excessively polluting private vehicles and recover transport costs from a cross-section of beneficiaries. For this, however, a number of experts have suggested the use of both public and private transportation with an optimum integrative interaction (Sriraman, 1997: 16).

Obviously, the focus needs to be placed on the issue of sustainability of both cities and transport. Rapid expansion of cities in India and a disproportionately limited resource base have already put excessive pressure on the existing public transport system. Basic improvement of public transport is thus urgently needed with pragmatic options. A Mass Rapid Transport System (MRTS), or Light Rail Transport (LRT) for all sizes of large cities with variations in *per capita* income needs to be re-examined. City size and the level of urban economic base also need to be linked up with public transport planning. Secondly, in order to be functionally viable, such systems simultaneously have to address the question of social equity. Experiences from different cities in low-income regions including Mumbai and Kolkata suggest that MRTS use rates are high only when fares are kept low through subsidies, when the city population density is high, when roads are congested, when there is an efficient feeder bus service and when private transport is costly (Saraf, 1998: 158). In such cases, the use rate, as mentioned earlier, is very high incorporating a large section from the higher income groups. Thus high capacity mass transit systems, in order to be economically viable, socially acceptable and environmentally sustainable, need to consider wider issues of urban and regional development.

Improving the existing public transport system is an optimal way to integrate the transport system with energy-efficient and environment-friendly mechanisms to suit the transportation needs of the urban majority. It also needs to be repeatedly argued that problems of urban transport cannot be sorted out in an isolated manner and have to be understood in a holistic way, considering the transport situation of the entire region. This would lead to the establishment of integrated transportation systems, made up of complementary transportation modes, compatible with the travel demand of, and affordability to, the majority of the urban population. With limited financial resources, capital intensive methods serving a fraction of the population should not be adopted. Instead, the system should be designed for achieving sustainability through an overall improvement in mobility, accessibility and safety.

Innovative methods need to be applied to increase the financial viability of the transport sector. BOT schemes have to be viewed in a correct perspective. Commercial exploitation of land adjoining mega-transport (freeway or expressway) projects can only be successful if the government's role becomes proactive in sharing related responsibilities. The suggestion of experts for the establishment of a 'Transport Development Bank' can also be considered. The latter would share the responsibility for establishing an urban public transport fund with contribution from various types of users such as property owners, employers, traders, and automobile owners. Such an effort may lead to an inter-sectoral equilibrium to allow a fair redistribution of resources, generated by the social benefits of public transportation. Accentuation of social disparity in the transport sector, due to the growth of personal motorized transport, will also be considerably mitigated by such efforts.

Developing a proper institutional base for achieving the above goals is imperative. The institutional framework would assure harmony among all networks, and would be capable of intervening in both public transport and the personalized automobile sector. A proper institutional framework actually would go further to allocate capital investment in infrastructure effectively and contribute towards viable land use policies to service the locational dynamics of future industries. To provide improved accessibility for a larger population as well as to protect the built and natural environment, the institutions should be strong enough to confront the demands of class elites and urge increasing use of public transport facilities by such groups. Only then can the social exclusion of the majority from transport infrastructure and related planning initiatives be reduced.

Sustainability programmes for urban transport in India, as well as in any country developing or developed, need to be a part of the sustainability programmes of the total urban system. Significant sustainable policy options at national levels thus have to be planned, based on principles of a sustainable society. Conceptual arguments and evidence from India indicate the existence of considerable confusion regarding 'sustainable city programmes'. While ample critique exists of the 'sustainable development' concept from the perspective of the South (Mahadevia, 2001: 21), confusion abounds on the concept of 'sustainable cities'. It is often promoted in the South as an environmental concept based on techno-managerial planning promoted by international development funding agencies that have found new avenues of funding that give recognition to the environment in relation to urban development. Such approaches towards 'sustainable cities' increase indebtedness and work towards the exclusion of the poor from the urban development process. It is exemplified by the recent hype about the new infrastructure like flyovers, expressways and freeways.

In order to make urban transport truly a part of the sustainable urban development programme, transport planning for the cities in India should be part of a wider planning process, having an incorporative approach to include the multidimensionality and multi-sectorality of development processes. In other words, sustainability and development should go side by side leading to an alternate, sustainable life style across the globe (White, 2000: 51). Essentially, the environmentally destructive practices of the richer countries need to be subsumed by the above choice that would help developing countries restrain themselves from copying the wrong practices. An incentive structure needs to be created for this that would encourage societies to re-use and recycle resources and limit the environmental degradation of one country, region and class by another. These directions alone can assure a sustainable future for the overall developmental initiatives in which transportation is an important element.

Notes

1. 'Road rage lobby' is a term commonly used in India to describe the roads lobby. It does not have the connotation of violent confrontation between motorists.
2. EMU – Electrical Multiple Units used for short distance suburban trains that have a structural system of three coaches.

11

Transport and Land Use in Chinese Cities: International Comparisons

Gang Hu

Introduction

With China's economy booming, questions have been raised about whether human life and environmental sustainability worldwide will suffer severe decline if China, the most populous country on the earth, were to increase its urban automobile ownership and usage to the level of the United States. A planner at one large automobile company believes that 'there could be 70 million motorcycles, 30 million lorries and 100 million cars in China by 2015' (Hook and Replogle, 1996). 'The potential effects of this car explosion – on the quality of human life and the sustainability of all life – are staggering' (Tunali, 1996: 4). If the Chinese were to drive as much as Americans, 'the carbon emissions from transportation in urban China alone would exceed 1 billion tons, roughly as much as released from all transportation worldwide today' (O'Meara, 1999: 143).

While not needing to review the problems of the automobile, which have been well-documented elsewhere (Gordon, 1991; Whitelegg, 1997; Newman and Kenworthy, 1999 and chapters in this volume) it is clear from the sheer magnitude of China's potential impact on the greenhouse effect and world oil supplies, that motorization within Chinese cities up to levels prevalent in the United States could be very threatening to the global environment. However, it is important also to consider the implications for China as a whole and particularly for life in Chinese cities. To understand the transport tendency in Chinese cities, it is necessary to explore the current situation of urban transport in Chinese cities. This chapter therefore investigates urban form, transport provision, transport patterns, and transport emissions in three major Chinese cities (Beijing, Shanghai and Guangzhou) and makes comparisons with cities internationally. The research of Newman and Kenworthy (1989) and Kenworthy and Laube (1999), have shown the value of such a comparison. Newman's, Kenworthy's and Laube's research provides the background and methodology for the comparisons between Chinese cities and other cities around the world.

Urban form

The population and employment densities and their geographical location in Chinese cities, particularly the relative location of homes and jobs compared with their Western developed counterparts, is of great significance for transport and environmental sustainability, as will be explained. The types of cities identified by Kenworthy and Laube (1999) are used below to make comparisons between Chinese cities and other cities in the developed and developing world. The types of cities are characterized as follows (Kenworthy and Newman, 1993, pp. 2–4):

- US City: US cities have experienced increases in car use per capita between 1980 and 1990. Despite some signs of resistance that are beginning to be seen in their transit systems, the automobile in US cities does appear to be 'unstoppable'.
- European City: European cities are managing to control automobile use much more successfully than in US cities. Transit use in many European cities has expanded, in some cities by more than one-third.
- Australian City: Australian cities are more like US cities in their continued patterns of automobile dependence, but car use levels are beginning to stabilize.
- Wealthy Asian City: Automobile dependence appears to be quite resistible in wealthy Asian cities, as indicated by the still very low levels of automobile dependence and large growth of transit in the modern Asian city.
- Other Developing Asian City: Other developing Asian cities are experiencing motorization. Due to high density and low transport provision, these cities have raised many transport problems though the automobile levels are still relatively low.

Population density comparisons

By international standards, Chinese cities, like cities in their Asian neighbouring countries, have high urban population densities and are characterised by quite intensively mixed land uses in their built-up areas. The average urban population density of Beijing, Shanghai, and Guangzhou in 1995 was 151 persons per hectare, which is three times that of the typical 'European city' (50 persons per hectare), and more than ten times the density of the typical 'US city' (14) and 'Australian city' (12). It was very close to the density of the 'wealthy Asian city' (153 persons per hectare) and the 'developing Asian city' (166).[1] The density of Shanghai (196) is lower than that of Hong Kong (301) and Seoul (245), but higher than most of other Asian cities.

The average population density in the central business districts (CBD) of the three Chinese mega-cities is 306 persons per hectare, which is four times that of the 'European city' (77.5 persons per hectare), six times that of the 'American city' (50), twenty two times that of the 'Australian city' (14), and even four

Table 11.1 Transport and urban form of the three Chinese cities

City	Urban density (person/ha)	Central city density (person/ha)	Urban area job density (jobs/ha)	Central city job density (jobs/ha)	Parking spaces per 1000 CBD workers	Length of road per person (m/person)	Average traffic speed (km/h)	Average total public transport speed (km/h)	Cars per 1000 people	Private passenger vehicle km per capita
Beijing	137.1	279.6	79.4	185.3	24	0.3	18.0	14.0	43	1141
Shanghai	195.7	227.3	130.6	150.2	3	0.3	16.0	12.0	15	743
Guangzhou	119.2	411.7	70.1	271.7	24	0.5	18.0	13.0	20	1314
Average	**150.7**	**306.2**	**93.4**	**202.4**	**17**	**0.4**	**17.0**	**13.0**	**26**	**1066**
US cities	14.2	50.0	8.1	429.9	468	6.8	51.1	27.8	604	11155
Australian cities	12.3	14.0	5.3	363.7	489	8.3	45.5	30.5	491	6571
Canadian cities	28.5	37.9	14.4	354.6	408	4.7	39.8	24.0	524	6551
European cities	49.9	77.5	31.5	345.1	230	2.4	33.4	37.2	392	4519
Wealthy Asian cities	152.8	86.6	87.5	881.9	80	1.8	27.5	30.7	123	1487
Developing Asian cities	166.4	281.9	65.1	279.3	192	0.7	23.8	16.8	102	1848

times the 'wealthy Asian city' (86.6). It is also higher than the density of other 'developing Asian cities' (282). The population densities in the Nanshi and Luwan districts of Shanghai are as high as 628 and 615 persons per hectare respectively. The Jin'an district of Shanghai and the Yuexiu district of Guangzhou also reach 531 and 510 persons per hectare respectively. These figures demonstrate that the urban form and land use pattern in the typical 'Chinese city' at present are quite different from Western cities, and much closer to their Asian neighbours.

Employment density comparisons

Employment densities in the three Chinese cities are also very high, with an average of 93 jobs per hectare in the urbanized area, which is three times the employment density of the 'European city' (32 jobs per hectare), eleven times the 'American city' (8), six times the typical 'Canadian city' (14), and eighteen times the 'Australian city' (5). It is also higher than most of the other Asian cities including Kuala Lumpur (22 jobs per hectare), Jakarta (59), Bangkok (62), Manila (68), and Singapore (49). Shanghai (131) can be compared with Hong Kong (140), which is the highest in the sample cities. The concentration of jobs in the CBDs in the three Chinese cities is, however, significantly lower than that of Western cities. The average CBD job density of the three Chinese cities is 202 jobs per hectare, which is half that of the typical 'American city' (430), lower than Australian, Canadian and European cities (around 350), and much lower than wealthy Asian cities (882).[2] Guangzhou (272) and Beijing (185) are close to the the 'other developing Asian' cluster including Jakarta (204), Manila (227), Kuala Lumpur (179).

The data presented above indicate that while *population* densities are typically high in the CBDs of Chinese cities, *employment* densities are generally lower than that of their Western counterparts. These differences reflect the characteristics of urban form and lifestyle in Chinese cities where both living and business activities are intensively focused in local areas having a broad range of mixed-use neighbourhoods.

Although many high-rise office buildings and large shopping centres have recently been built in the central areas of Chinese cities, unlike in Western cities, centralization has not been accompanied by the significant dispersal of population. The population densities both in central areas and the wider urbanized areas remain very high. US and Australian cities, on the other hand, have very low population densities but high job densities in their central areas, which means that working and living places are geographically strongly separated. This is the major reason for the high traffic demand in such cities as we will see in later comparisons.

These data suggest that the typical Chinese city in common with other developing Asian cities exhibits a mixed residential and commercial land-use without a strong and distinct central business district as in Western cities. This is compatible with the findings of Rimmer (1986) that developing Asian cities appear to have weak centres of business activity, because they have been developed from the 'walking city' through motorization, mainly based upon bus transport, in contrast with most cities of Western countries which grew around radial rail systems which helped focus jobs in a single centre. Lower capacity and slower bus systems are not nearly as effective at concentrating jobs as suburban railways. Today in the developed world mixed land-use is seen to be preferable as it enriches working and living environments, reduces vehicular trip-making, and offers more opportunities for walking and cycling to work (Cervero, 1986; Whitelegg, 1997).

The difference in land use patterns between Chinese and Western cities reflects the legacy of several decades of urban development under strict government control in the absence of a land market, especially the widespread implementation of enterprise-based provision of housing and other services. It is true that with new urban development strategies and the emergence of an urban land market that makes land values reflect locational advantages, some changes have been taking place. These include the large-scale addition of office buildings in the city centre, relocation of manufacturing businesses from city centre to the suburbs, and new development of residential zones in the city fringe areas. Nevertheless, the basic pattern of urban form and structure is maintained and one would not expect significant change in the foreseeable future.

Transport provision

Road infrastructure

One of the most important transport infrastructure indicators is the amount of roadspace in the city. The length of road per person in Beijing is 0.3 metres, in Shanghai also 0.3 metres, and Guangzhou 0.5 metres per person. The average for Chinese cities is one-seventeenth of the US cities (6.8 metres per person), one-sixth of European cities (2.4) less than one-twentieth of Australian cities (8.3). The figures for Chinese cities are close to Hongkong (0.3), Surabaya (0.3), Jakarta (0.5), and Manila (0.6). The high automobile-dependent and low density cities in the USA and Australia have high provision of road space (7 to 8 metres per person), Canadian cities have somewhat lower provision, while European and Asian cities show lower levels still – down to 0.7 metres per person.

As roads are also the mechanism by which other basic urban infrastructure facilities such as water, power, and sewerage are delivered to properties, road length per person in some ways measures the efficiency with which these other items are supplied in different cities. This of course varies strongly with density, with lower density cities, characterized by sprawling development, needing to provide much greater length of reticulated infrastructure than more compact environments.

Parking provision in the central city

Another important determinant of transport patterns is how much parking space is provided. This is particularly important in the critical central city area or CBD where space is always most constrained and where it becomes especially clear where the city's priorities lie in relation to transport.

In Chinese cities, on-street parking is very restricted on most major roads, particularly during peak hours, but it is available on many minor roads. Due to the very high price of central city land, off-street parking in the CBDs of Chinese cities is actually very limited, and most car parks are working units owned by proprietors for their own vehicles. As with other goods and services, parking pricing is strictly controlled by government (through the Bureau of Pricing). The government permitted parking rates are low compared with the high value of CBD land. The operation of parking lots is not rewarded. Parking fees for a car in Beijing in 1995 were US$0.12 for four hours.

According to the *Shanghai Central Area Motor Vehicle Parking Management and Strategy Research Report*, there were only 1,082 off-street parking spaces in the central area of Shanghai, and the total on-street and off-street parking in the Shanghai CBD was a mere 10,468 spaces. The report reveals that, although the demand for parking is much greater than the space available, still some parking lots under high-rise buildings were actually renovated for retail shops because greater profits could be had from the lease for that purpose. Chen (1998) reveals that the demand for motor vehicle parking at 13.00 (1.00 pm) on an average day in the Shanghai CBD was for 18,637 spaces. As a result, it is not surprising to see that some motor vehicles park on the footpath, in non-vehicle lanes, in non-permitted on-street spaces, and in public activity areas. In an international comparison, parking spaces per 1000 CBD workers in the three Chinese cities were extremely low at 17 spaces per 1,000 CBD workers. This compares with 468 per 1,000 in the typical US city and 230 in the European City.

CBD parking is of course a critical factor in helping to determine the modal split for the journey-to-work to the CBD. If parking is physically not available, then clearly other modes must cater for travel demand. This is a much more critical factor than pricing. Experience suggests that there always tends to be some segment of the travel market prepared to pay whatever price is set for the available parking supply (TEST, 1989). Parking policy in Chinese cities is unique. Due to the very high demand for public parking spaces in the central city area, the Chinese government is encouraging construction of parking lots under high-rise buildings. But, on the other hand, the pricing system actually discourages the operation of parking lots as a business. This inconsistency makes the outcome uncertain in terms of transport priorities.

Congestion and travel speeds

One of the key outcomes of the different patterns of transport infrastructure in cities is the speed of the traffic system (reflecting congestion levels) and the speed achieved by public transport modes. The relative speed of private vehicle traffic as compared with the speed of public transport is a basic factor in helping to determine travellers' modal choice (Kenworthy and Laube, 1999). The data demonstrate that both the average vehicle traffic speed (17.0 kilometres per hour) and the average public transport speed (13.0 kph) of the three Chinese cities are significantly lower than all the other international cities in the sample except Bangkok (13.1 and 9.2 kph respectively). It is important to understand that the three Chinese cities have experienced the highest congestion levels in the sample of cities. However, the ratio of transit to car speeds of the Chinese cities (0.76) is higher than that of US (0.55), Australian (0.67), and Canadian (0.60) cities. But it is lower than European and wealthy Asian cities (1.12 and 1.15 respectively), and close to developing Asian cities (0.71).

Transport patterns

Vehicle ownership and private car use

Car ownership in most Chinese cities is still very low: Beijing (43 cars per 1,000 persons), Shanghai (15), and Guangzhou (20). The average number of cars per 1,000 people of the three Chinese cities is only 26 compared with 604 for US cities and 123 for wealthy Asian cities. This compares with the 'US city' (604), the 'wealthy Asian city' (123) and 'developing Asian city' (102). Because there is still severe congestion even with such low vehicle ownership it appears that Chinese cities do not have much space to accommodate automobile use.

The other important variable describing automobile dependence is the passenger kilometres travelled per person in private passenger vehicles. Chinese cities (1,066 km per person) are also very low in private car use. The figure is only one-tenth of that of the United States (11,155), and lower than that of both wealthy Asian cities (1,487) and developing Asian cities (1,848). It is notable that per capita levels of private mobility in US cities are in a league of their own. The per capita private transport mobility in the USA is even higher than transit and private mobility combined in all other cities. This helps explain why urban dispersal and road building result in greatly increased travel demand. In the three Chinese cities, by contrast, the average journey to work length is as follows: walking 1.3 kms, cycling 4.7 kms, and bus 8.7 kms. These three transport modes account for 92.5 per cent of all trips. Even the average trip length by car in the Chinese cities is only 11.2 kms, which is much lower than in Western cities (see Table 11.2).

One of the important outcomes of the private transport patterns discussed above is the amount of energy used to support the system. The 'Chinese city' consumes, on average, 2,338 megajoules per person per year. As with other private transport indicators, the 'Chinese city' is substantially low compared

Table 11.2 Transport modes and pollution levels of the three Chinese cities in international comparison

City	Private passenger energy use (MJ per capita)	Transit passenger trips per person	% of workers using transit	% of workers using private transport	% of workers using foot or bicycle	Total transport CO_2 per capita (kg)	NO_x per capita (kg)	CO per capita (kg)	VHC per capita (kg)
Beijing	3333	456	32.4	6.1	61.5	400	11.5	53.5	14.1
Shanghai	1567	452	15.1	7.0	77.9	201	8.5	39.7	10.5
Guangzhou	2114	215	21.2	9.4	69.4	286	17.0	79.4	20.9
Average	**2338**	**374**	**22.9**	**7.5**	**69.6**	**296**	**12.3**	**57.5**	**15.2**
US cities	55807	63	9.0	86.3	4.6	4541	22.3	204.5	22.3
Australian cities	33562	92	14.5	80.4	5.1	2811	21.9	185.8	23.0
Canadian cities	30893	161	19.7	74.1	6.2	2434	27.0	160.6	21.7
European cities	17218	318	38.8	42.8	18.4	1887	13.0	72.6	11.6
Wealthy Asian cities	7268	496	59.6	20.1	20.3	1158	6.2	19.8	2.2
Developing Asian cities	6819	334	37.8	43.9	18.4	837	8.7	61.8	13.6

with other cities in the sample, especially the 'US city'. Every 24 Chinese urban dwellers consume the same energy in private transport as one US urban resident does. In this sense, then, Chinese cities are currently much more sustainable than US and even European cities.

Public transport service and use

Data show that although transit provision in Chinese cities is far from sufficient, there is relatively high transit use. Transit vehicle kilometres of service per person in Chinese cities is only 38, second lowest after the USA (28), much lower than Canadian (58), Australian (60), European (92), developing Asian (108), and wealthy Asian cities (114). The length of railway (metres per 1,000 persons) in Chinese cities is by far the lowest in the sample. However, the average annual transit passenger trips per person of the three Chinese cities is 374, which means an average of about one trip on transit per day for every man, woman and child. The figures for Beijing (456) and Shanghai (452) are higher than for Guangzhou, and similar to Singapore, Tokyo, Seoul, and Manila, but lower than Hong Kong and Zürich. All three Chinese cities are significantly higher than the US, Australian or Canadian cities. The average trip length of transit in Chinese cities, however, is shorter, since they benefit from high density and mixed land use, especially in the central city area.

Certainly transit plays a more important role in the 'Chinese city' than in both North American and Australian cities. It is especially significant that annual passenger-kilometres per vehicle and annual boardings per vehicle are extremely high in Chinese cities. Passenger kilometres per vehicle are 6.4 million in Beijing, 2.5 million in Shanghai, and 1.5 million in Guangzhou in 1995, while the average speed is only between 12 to 15 kilometres per hour (Hu and Kenworthy, 1999). The average passenger usage per vehicle in the three Chinese cities is 3.5 million passenger-kilometres per vehicle. In an international sample of 13 cities in 1995 the average was 1.8 million passenger-kilometres per vehicle (unpublished 1995 data collected for the UITP database by Kenworthy and his group).[3] Likewise, the overall average speed for public transport (all modes) was 30.5 km per hour or approximately double the average speed of the Chinese bus system (unpublished 1995 data collected for the UITP database by Kenworthy and his group). This is consistent with the crowded situation in buses in most Chinese cities. Although the bus fleet in 640 Chinese cities doubled in five years up to 1995, it is still far from satisfying the demand. It is reported that passenger density on buses sometimes reach thirteen people per square metre in peak hours (Wang, 1995). The major result of poor transit provision is that transit usage is depressed, which means the percentage of total passenger kilometres on public transport could be much higher if the services were improved to the levels found in Hong Kong, Singapore, and some European cities. Kenworthy and Laube (1999) points out that the significant difference between wealthy Asian and developing Asian cities in most major transport characteristics can be attributed at least in part to a gap between supply and demand in rail transport in the latter cities.

The provision of fixed rail transport in the Chinese cities is extremely low. Bus and trolley bus are the predominant transit modes. Only 7 per cent of transit travel is by rail mode in Chinese cities, compared with 77 per cent for the European city, and 57 per cent for the wealthy Asian city. Even in US cities the proportion reaches 32 per cent. What this suggests is that Chinese cities are ill-prepared in terms of public transport development for the onslaught of the private car (see also Chapter 15 of this volume). Trips could tend to shift from non-motorized modes to cars and motorcycles if transit systems are not better developed.

Modal split in the journey to work

As stated above, another opportunity presented by the high urban densities is the possibility that many trips can be short and therefore easily made on foot or by non-motorized vehicles. Mixed land use encourages short trips and non-motorized transport by allowing a diversity of destinations to be available within a short distance. Data demonstrate that walking and cycling are still the most popular modes of travel in the Chinese city. Cycling accounts for 49.7 per cent, 45.1 per cent, and 34.4 per cent and walking 11.8 per cent, 32.8 per cent, 35.0 per cent in Beijing, Shanghai and Guangzhou respectively. The total percentage of workers using non-motorized modes averaged 69.6 per cent in the three Chinese cities, compared with 4.6 per cent in US cities, slightly higher in Australian and Canadian cities, 18.4 per cent in both European and developing Asian cities (for very different reasons), and 20.3 per cent in wealthy Asian cities.

By contrast, the use of private motorized modes in the Chinese city is extremely low, which is consistent with lower car ownership and use. The average percentage of workers using private transport is only 7.5, compared with 86.3 in US cities, 80.4 in Australian, 74.1 in Canadian, 42.8 in European, 20.1 in wealthy Asian, and 43.9 in developing Asian cities. It is important to notice that the 'Chinese city' is distinct from the other developing Asian cities in the Kenworthy and Laube sample in which more use is made of motorized private modes.

Transport emissions

Global and local impacts of transport emissions

Urban transport contributes significantly to a number of important human impacts on the global environment. Prominent among these is the build-up of greenhouse gases, in particular carbon dioxide (CO_2), in the atmosphere. Data show that Chinese cities currently contribute very little *per capita* to the global greenhouse effect compared with other cities around the world. The annual average transport CO_2 emission per capita of the three Chinese cities discussed is only 296 kg. An average Chinese urban dweller contributes only 6.5 per cent of the average American in CO_2 emissions from total transport and 25.6 per cent of the average European. This also compares with 25.6 per cent of the average person in wealthy Asian cities. If every person in the world is entitled to take equal responsibility for global warming (as is argued by Baer *et al.*, 2000), the populations of American and other Western cities are much more responsible for global warming than the people of Chinese and other Asian cities.

In terms of local impacts, nitrogen oxides (NO_x), carbon monoxide (CO), volatile hydrocarbons (VHC), sulphur dioxide (SO_2), and particulates (SPM) are the major noxious pollutants. Unfortunately, data for the latter two are not available for Chinese cities. For local impacts, because population densities are so high it is important to focus on spatial exposure. In the case of polluted air, having more people exposed to air pollution within a defined area does not mean that each person inhales less polluted air. So spatially based indicators better explain the intensity of pollution than per capita ones. The *per hectare* figures for air pollution in Chinese cities are very high, due to the high density of urban land use. The average annual NO_x, CO, and VHC emissions of the three Chinese cities are 2,196 kg, 10,238 kg, and 2,695 kg per hectare per annum for Beijing, Shanghai and Guangzhou respectively. These are much higher than all the other cities in the sample.

Emissions from the transport sector in Chinese cities at present constitute a substantial threat to the health of their local populations. By contrast the contribution of these cities on a per capita basis to global warming is much lower than the contribution from cities of the developed world. The local threat may therefore act to limit the development of Chinese cities along the lines of the American or even the European model with high levels of automobile dependence. This may counteract to some degree the potential global threat of carbon emissions from rapid urbanization in China

Analysis and suggestions

Urban density

The limitation of available land for urban development does not allow Chinese cities to follow the US urban form, which is characterized by very low density with urban sprawl. Nevertheless, experiences in some other developing Asian cities such as Bangkok, Jakarta and Manila have demonstrated that shortage of land certainly does not mean that urban sprawl will not occur. Although high population density has generally been retained in the central areas, rapid development at the urban fringes has resulted in severe decentralization in Bangkok, Jakarta and Manila in the last ten years. Will this scenario occur in Chinese cities?

The answer is still uncertain. However, although criticisms can be made of China's political system, it is centralised and the government has power to implement its policies. China has a reasonable planning system which provides the possibility that urban development can be controlled effectively. The current urban planning and land use laws emphasize retention of high density and prevention of urban sprawl. If these current policies continue in force, and other economic and social policies such as the 'automobile industry policy' can be made consistent with them, then Chinese cities should be able to take a different path from that of the other developing Asian cities.

Chinese cities are the most congested and air-polluted of all cities in the international sample considered by Kenworthy and Laube (1999) at a time when motorization is still at a very low level. Although other factors such as traffic management and motor technology also contribute to these issues, it is obvious that the extremely high density with a large population centralized in a relatively restricted urban fabric is the major cause of pollution. High density means that motor vehicles encounter substantially more intersections for the same length of trip. This causes lower average traffic speeds and, importantly, more stop–start actions, both resulting in significantly increased emissions.

Poor management of intersections and the mixture between motorized and non-motorized transport in the road system also result in lower traffic speed and more stops and starts for motor vehicles, which cause increased emissions. The motor technology of locally manufactured motor vehicles is also one of the reasons that the average emissions per kilometre of travel of motorized vehicles are high in Chinese cities. Nevertheless, these are obviously minor factors compared with the urban form. Beijing, for example, has required the installation in all vehicles of a device which helps to reduce emissions, and has required all taxis to use liquid gas since 1996. However, the overall air quality in Beijing has not improved significantly.

Although the average kilometres of motor transport per capita is very low in Chinese cities, because of the high population and employment density, the motor trips are concentrated in a relatively small urban area. Therefore, the motor traffic intensity (or average kms of motor travel per hectare) is comparable with Western cities. Furthermore, for the above reasons, the amount of emissions is much higher for a trip of the same length in a high density area. It is thus easy to understand why air pollution has been severe in Chinese cities. In this sense, automobile dependence is more harmful locally in high density cities, as a small increase in automobile use will create serious local environmental problems.

It is estimated that in 1995 trucks (lorries) contributed approximately 36.6 per cent of total emissions in Beijing and 33.3 per cent in Shanghai.

Buses contributed approximately 12.2 per cent in Beijing and 31.4 per cent in Shanghai. These figures also demonstrate that, in contrast to Western cities, automobiles are not the overwhelming factor in the total transport emissions in Chinese cities, and that trucks and buses alone could cause serious air pollution in high density cities.

It is important, therefore, to develop multiple suburban sub-centres for large high density cities to reduce the average travel length and therefore minimize total travel demand. The practice in some Western cities shows that the sub-centre strategy, to some degree, generates new auto traffic demand, as the sub-centres usually provide large numbers of parking spaces which actually encourage auto use. The current practice in large Chinese cities has not raised this issue, because parking space remains in relatively short supply. However, it should be noted that the provision of additional parking space to meet demand could create a potential problem in physical planning in the future. It is also important to note that electric-powered public transport should take the major share of travel in large high density cities.

It is understandable that worries about greenhouse gases emitted from motorisation in Chinese cities stem from the experience of heavy air pollution in Chinese cities. This study demonstrates that the severe air pollution is mainly due to the high density of occupation, and that average emissions per capita are very low in Chinese cities in comparison with Western cities. It is therefore unjust for Western countries to worry so much about Chinese cities destroying the global environment while their own citizens contribute ten times as much greenhouse gases. A large urban population by itself is not a reason for criticism.

Transport modes

As Thomson stated in his book, *Great Cities and Their Traffic,* the car in itself is not a devil, it is a convenient means of private transport when it is appropriately used. The problem for Chinese cities, and indeed all cities, is how to deal with the issues raised by mass car ownership and increasing dependence on the automobile to the excessive detriment of local air quality and other transport modes.

China cannot afford the huge consumption of arable land and crude oil that massive automobile use requires. Severe local air pollution at a time when automobile traffic still constitutes a very low percentage of total transport also suggests that Chinese cities simply could not survive if they were to reach the automobile levels found in American cities. Automobile transport is therefore not a viable solution for Chinese cities. The question, however, is: to what degree should Chinese cities control their auto ownership and usage? There are some options such as European cities' medium automobile ownership and medium usage, Singapore's low automobile ownership and high usage, or Hong Kong's low automobile ownership and low usage. The European model is more suitable than the American and Australian models. But European cities are based on medium densities. Considering the limitation of available land, the European model is also not suitable for China. As to the developing Asian cities, the scenario of Bangkok's transport development is obviously not the one that Chinese cities would want to follow.

The Singapore model is often recommended by Western commentators. The data of Kenworthy and Laube (1999) show that automobile ownership in Singapore is 101 cars per 1,000 persons, which is fairly low in the international sample of 46 cities. Nevertheless, kilometres travelled per car in Singapore is 18,370 km, which is around 1.4 times of that of Sydney, Melbourne, Brisbane, and Perth. Automobile *use* in Singapore is thus rather high, higher even than in the Australian cities. The Singapore solution strictly controls automobile ownership through pricing measures, but it provides a high quality road system and a massive amount of parking space which actually encourage car use. The reasons for Singapore's success are: that it is a semi-island city surrounded by a water body, so motor vehicle emissions are easily blown away; public transport is mainly electric powered; and the overall urban density is relatively low (87 persons per hectare, compared with an average 151 persons per hectare of Chinese cities). The large Chinese cities do not have the natural advantages that Singapore has. The large Chinese cities also could not reach a density as low as that of Singapore. As stated above, because of high densities and concentrations of motor vehicle emissions, Chinese cities are suffering severe air pollution even when both automobile ownership and usage are significantly lower than that of Singapore. In this sense, Singapore's model is also not the one that Chinese cities should follow.

It is obvious, then, that the only option for Chinese cities is low automobile ownership and low usage. To comply with this low auto transport strategy, it is important to provide a well developed system for other transport modes. Transit incentives are another key ingredient in policy to help control cars and balance urban transport.

The major factor in the current low car ownership and usage in Chinese cities is the level of income. With a continuous high rate of economic development, the number of high income families will greatly increase in the coming decades. This change may translate into a stronger motivation for car purchasing and using. It is desperately important, therefore, to provide comfortable, convenient, and high speed transit systems to attract potential car users while implementing car *disincentive* policies. And it is desperately necessary to provide durable, economical, and high-capacity transit systems on rights-of-way separated from street congestion. The sustainability agenda also demands transit systems that are competitive with the car in passenger appeal and speed. High densities create sufficient concentrations of activity for a very effective, frequent public transport service. Despite this positive land use framework in Chinese cities, providing better public transport is actually a key issue in Chinese urban transport. Furthermore, with high density and traditionally large shares of walking and cycling, incentives for these modes too are particularly important in Chinese urban transport planning (see also the case of India discussed in Chapter 10).

Road widening and construction of multi-level intersections for increasing motor vehicle use exposes pedestrians and cyclists more and more to the dust, motor vehicle emissions, and noise of horns. In Chinese cities it is becoming more and more difficult to cross the road in many places, and sometimes one has to travel an extra 3–5 kilometres via a multi-level intersection to reach a destination which was only two hundred metres away. The perception of safety and convenience of non-motorized travel is declining in many Chinese cities, while studies demonstrate that 'perceptions are powerful' (Roseland, 1998). To maintain non-motorized modes and encourage more people to walk and cycle, networks of pedestrian and bicycle routes that are perceived as safe and convenient must be maintained and developed.

While walking and cycling are recognized as 'green transport', they also have their weaknesses on rainy or windy days and under some geographical conditions, especially when longer journeys are necessary. Integrating walking, cycling and public transport is a feasible and effective strategy. Those who travel by bus or subway can cut their travel time considerably by going to the bus stop or station by bicycle.

Bus and rail systems

Data showed that transit usage in Chinese cities is at a medium level. Nevertheless, the density of railways in Chinese cities (2.8 metres per 1,000 persons and 428 metres per 1,000 hectares) is extremely low compared with those of European cities (261.3 m/1,000 persons and 11,536 m/1,000 ha), wealthy Asian cities (38.0 m/1,000 persons and 4451 m/1,000 ha respectively), and even the developing Asian cities (10.4 m/1,000 persons and 1,791 m/1,000 ha respectively). Bus boardings account for 94.9 per cent of the total public transport and bus passenger kilometres account for 88.9 per cent of the total public transport in Chinese cities. These figures demonstrate that the transit system relies heavily on the bus fleet, and railway systems are totally undeveloped in Chinese cities.

A small further increase in motor vehicle use could cause big problems of air pollution in Chinese cities. As buses contributed approximately 12.2 per cent of the total emissions in Beijing and 31.4 per cent in Shanghai (in 1995), it is obvious that the development of public transport in these Chinese cities should not rely on carbon fuelled bus systems which would greatly increase emissions. A rail system is faster and has much larger carrying capacity than a bus system, and therefore it is more attractive and consumes less urban land. A rail system is flexible and adaptable since its carrying capacity can be adjusted by changing the vehicle units according to the demand for different service hours, and therefore it is economic. Electric railways do not contribute to urban air pollution. The problem of developing a rail system is that it requires a huge amount of investment. Nevertheless, such investment would be practical if more funds could be transferred from road construction and automobile levies. And the investment could partly expect to be returned through comprehensive developments with high value retail and residential apartments around the rail stations.

Rail systems should form the main skeleton of the public transport network in Chinese cities, while feeder bus systems are still needed to play their role in the lower level of the public transport hierarchy. Beijing has been extending its rail network. Shanghai and Guangzhou have started the construction of their rail systems in recent years. These developments demonstrate that rail systems are feasible and practical in Chinese cities.

Conclusions

The growing trend towards motorization in Chinese cities presents a number of important opportunities and challenges, both for the cities themselves and potentially for the global environment.

For Chinese cities, the automobile means a whole new way of life for the residents of the world's most populous country, offering levels of personal mobility previously unthinkable. For global car manufacturers, China offers one of the biggest markets in the world. The benefits will, however, exact a terrible toll on the urban environment and the quality of public spaces and human interaction in Chinese cities, as the automobile has in countless other cities. Accommodating the automobile will also have a big impact on China's land supply for agriculture and on its economy through a whole range of new costs.

Chinese cities, however, need to be understood in an international context before the spectre of unbridled automobile dependence, as exhibited in the US, is accepted uncritically as a reality for the future. When a detailed analysis of Chinese cities is carried out, the data reveal urban environments of very high density and mixed use urban forms. Analysis also reveals levels of car ownership and use that are, and will be for the foreseeable future, very low by world standards. Non-motorized transport, though under increasing threat, is still very strong, and public transport, by international standards, is still reasonably healthy in usage, though significantly constrained in its fleet capacity and service levels.

In any country, mobility demands present many conflicts and obstacles. There is no doubt that China is at the crossroads in dealing with its urban transport systems. Nevertheless China can call on a long tradition of government planning and stewardship in its efforts to balance the positives and negatives of a trend towards increasing car ownership and use. In this context, the existing massive and growing contribution to global greenhouse gases and world oil consumption made by the developed world is presently more of a threat to global survival than the potential motorization of China.

Notes

1. Data for international cities are from Kenworthy and Laube (1999). Data for the Chinese cities are from various government agencies, research institutes, and publications (Hu, 2002).
2. It should be noted, however, that in the case of CBD employment density, for different reasons some cities lie well outside the clusters: Phoenix (USA) records only 90 jobs per hectare and Sacramento (USA) 117. Amsterdam in the Netherlands has only 98 jobs per hectare.
3. The cities were: Osaka, Sydney, San Francisco, Helsinki, Atlanta, Tokyo, Berlin, Hong Kong, Manila, Melbourne, Oslo, Perth and Wellington. Some cities had less than 0.5 million passenger/km per vehicle.

12
Barriers to Sustainable Transport in Australia

Brendan Gleeson, Carey Curtis and Nicholas Low

Introduction

Popular imagery often celebrates Australia as a paradise of sweeping deserts and pastoral plains, ringed with unsullied golden beaches. The image is a deeply misleading one for two principal reasons. First, Australia, the nation, is a thoroughly urban society whose peoples have largely abandoned rural and outback living for city life. Two thirds of Australians live in the nation's eight largest cities and nearly 4 in 10 citizens live in Sydney and Melbourne alone. The suburban bungalow and the private motor car, not the distant farm and the long journeying train, are the principal features of Australian life.

Second, the myth of 'clean and green' Australia, celebrated most recently with Sydney's 'Green Games', masks a far more complicated environmental reality that we cannot here survey in detail.[1] Suffice it to say that, on many indicators, Australia and Australians have overreached the limits of sustainability. Most alarming is the fact that Australians are now the highest emitters of greenhouse gases in the world (tonnes of greenhouse gas per capita, Miller, 1999:3).[2] Compared to other advanced capitalist nations, Australia has a relatively small heavy industry sector. The origins of the greenhouse 'blowout' are manifold, but include the nation's obsessive reliance on the private motor vehicle and its prolonged major investment in the infrastructure that permits this obsession to continue. In recent years a panoply of new major road projects has been undertaken by regional (State) governments, urged on, and partly financed, by a federal administration that celebrates the 'economic stimulus' that such works allegedly deliver. Within key state and private institutional domains, the 'road machine' grinds relentlessly on, underscored by the logic of economic responsibility. In these straitened and competitive global times, it would be nothing less than irresponsible for regional governments to deny their aspiring capital cities the investments that would permit the free flow of goods and services across their cluttered urban surfaces. Or so the story goes...

Reformist pressures and their consequences have been evident for decades in Australia. From the early 1970s, urban social movements emerged to oppose giant freeway building plans. The political heat generated by these eruptions of discontent caused some State governments to defer (though rarely to tear up) freeway building plans. Three decades on, however, it is apparent that, after moments of subordination to alternative policy settings, conventional car-dominated transport planning has reasserted its pre-eminence (Laird *et al.*, 2001). Key road building institutions have proved to be extraordinarily resourceful in adapting to changing institutional and political–economic circumstances (notably 'globalization' as interpreted by neo-liberalism) from which they have drawn new power. 'Sustainability' has been reduced to one of a number of competing criteria whose claims are balanced against one another in benefit–cost analysis. If a transport solution is unsustainable it can nevertheless safely be pursued so long as it improves the chances of economic growth. Thus the discourse of liberalism neutralises that of sustainability.

This chapter reviews the recent history of transport planning in three Australian cities, Sydney, Melbourne, and Perth, enabling us to chart how transport discourses are constructed and contested at the metropolitan scale. Sydney, Melbourne and Perth are the metropolitan capitals of the States of New South Wales, Victoria and Western Australia respectively. The Australian experience of transport planning is by no means homogeneous. Our analysis highlights the ambiguities and periodic reversals which have attended the shift towards mass transit in Sydney and the continuing strength of the roads bureaucracy and road-based planning in Melbourne and Perth. From these empirical analyses we draw conclusions about the barriers to transport sustainability and the path of ecosocialisation that was mapped out in Chapter 1.

Forwards and backwards in Australia

The Australian nation recently celebrated the centenary of its existence.[3] And yet, a century after its formation, Australia still has no national rail network, and federal expenditure on transport infrastructure has in the past overwhelmingly favoured roads for reasons similar to those supporting such funding in the USA. Government spends money on roads not just because it is good for business in the short term (it is) but because it is popularly approved. Between 1975 and 2000 the Commonwealth (federal) Government spent about AUS$40 billion (about US$26 billion) on roads compared with about AUS$1.9 billion (about US$1.2 billion) on railways (Davidson, 2000: 15).

The 2000 budget forecasts allocate AUS$71 million to rail in 2000/01, contracting to zero by 2003/4. By contrast, AUS$2.9 billion will be provided in grants to the States to be spent on roads between 2000 and 2003 (Commonwealth Government, 2000: 48 and *passim*). As in the USA, this enormous pot of federal money for roads is unquestionably part of the explanation for the enthusiasm of State governments for road building. If the money earmarked for roads is available, it is politically difficult for state politicians to turn it down.

But powerfully underscoring the allure of federal money is the simple ideology of 'growth by asphalt' which celebrates road building as principal motive force in economic growth. For example, the Allen Consulting Group in 1993 estimated that an extra billion dollars invested in urban freeways would increase GDP by 0.15 per cent over a 10-year period (O'Connor *et al.*, 2002: 186). Such calculations always defer and disperse the real price to be paid.

While the Commonwealth Government provides a large proportion of the funding, provision of transport infrastructure is the responsibility of State governments and is partly paid for by state taxes. While, federal funds are raised from taxes on fuel and are therefore paid 'at the pump', state taxes from motor vehicles are paid annually in block form (similarly, the cost of motor vehicle insurance is met by a single 'up front' annual payment). Payment for public transport, on the other hand, is mostly made per trip at the point of the sale. Once these payments are made, the price of an additional journey seems less to the traveller than that of a trip by public transport, so there is a bias in the pricing mechanism in favour of roads.

The ideal of a free flow of road traffic which would cut the costs of congestion was transmitted to Australia from the USA by a continual two way traffic of engineers. The road engineers were educated in America. The Australian transport research organizations were set up on American lines. American consultants provided a stream of advice on the benefits of planning freeways as the solution to urban congestion. These themes – unbalanced federal funding, 'growth by asphalt', the free flow utopia disseminated from America, and biased pricing emerge continually in the infratsructure planning of Sydney, Melbourne and Perth.

Sydney, 'The accidental city'

Ashton's (1995) historical survey of planning in the City of Sydney casts Australia's principal metropolis as an 'accidental city' that emerged through a series of spontaneous, organic and largely uncoordinated developments. Planning of a sort was nevertheless evident (see Spearritt and Demarco, 1988), and amongst the various players in postwar urban governance in Sydney, it is the road building agencies that have had the most success in terms of their own objectives.

As Searle (1997, 1999) explains, the Department of Main Roads (DMR) imported and began to sell the American vision of the 'efficient freeway city' before the Second World War. Historical geographers speak of 'diffusion' of ideas and practices across space, emanating from a core. Diffusion occurs through mechanisms such as 'migration chains' in which inter-personal connections provide the basis for a wholesale or partial generalization of cultural forms. In our theoretical schema, a discourse, such as the freeway ideal, may diffuse and strengthen – eventually perhaps into a discursive regime that governs broad sets of practices – through subtle and often unremarkable chains of interpersonal contacts. There is no doubt, for example, that the freeway ideal diffused to Australia from the USA in this way through both personal contacts between key road agency staff and the actual migration of individual professionals for short and longer periods between both countries (Mees, 2000).

The Main Roads Development Plan (MRDP) conceived between 1938 and 1946 was 'the first long term plan for freeways and other major urban roads in Australian cities' (Searle, 1997, p. 3). The plan's familiar solution to projected sclerosis of the city's inner traffic arteries involved construction of a system of radial freeways focusing on the city centre. The MDRP provided an important foundation for Sydney's first regional plan, the Cumberland County Plan in 1948 (Spearritt and DeMarco, 1988). In the years that followed, financial prudence, of a sort, became the hallmark of transport planning. 'Inefficient and outdated' public transport systems were to some extent run down and, in most cases starved of investment, as the discourse of freeway modernization took root in the political–institutional cultures of postwar State governments.

The extensive tramway system was dismantled between 1950 and 1961 and only partially replaced by public buses (Spearritt and DeMarco, 1988). Meanwhile, radial corridors of land for the planned freeways were carefully purchased beyond the urban fringe, thus securing the basis for further network expansion and low density urbanisation. This prudence in part explains the enduring power of the road agencies and the lasting reach of their visions. 'By the time the government turned to private financing of expressways in the late 1980s, it was usually able to offer routes which had already been purchased or, at the least, been reserved by zoning since the 1951 plan' (Searle, 1997, p. 3).

The DMR remained a central force in urban governance through to the 1970s. The 1968 Sydney Regional Outline Plan (SROP), ostensibly a product of the State Planning Agency (SPA), reflected this fact, retaining the radial expressway system (Searle, 1997, p. 3). The projected increase in central Sydney employment, which underscored the freeway plan, proved baseless when the 1966 national census results were published and analysed (ibid.). No matter, Australian road agencies have rarely been troubled by regular exposure of their (sometimes spectacular) miscalculations of putative demand for 'road products' (see Beed, 1981, and Mees, 2000 on this).

The first serious counterdiscourse to trouble the freeway vision for Sydney emerged in the late 1960s. Grassroots opposition to freeway (then 'expressway') building surfaced in the newly gentrifying inner city suburbs, notably Glebe. Developers and bureaucrats condemned such opposition as anti-modern.

But in a unique class alliance freeways were subjected to the 'green bans' imposed against environmentally damaging development by organized labour, principally, the state Builders Labourers' Federation (BLF), in collaboration with local environmental activists. The first anti-expressway green ban was imposed in Sydney as early as 1962 (Burgmann and Burgmann, 1998.). The Green Bans were the first, and perhaps still the most forceful, expression of a new alignment between social protection and environmental conservation. This struggle, like many others subsequently, was over *local* environmental conservation, not about the global ecological threat posed by transport.

The position taken by sections of organized labour influenced the transport policy of a new Federal Labor government elected in 1972. Federal support for the DMR's plans was greatly reduced. The 1977 election of a new State Labor government amidst entrenched opposition to freeway building saw the DMR's inner urban freeway plan finally abandoned. The new State government also undertook some badly needed recapitalization of urban public transport, including the railways. A new Eastern Suburbs line was opened in 1979 amidst dire predictions that the project would quickly fail. However, the line was soon attracting patronage that was 25 percent in excess of original project forecasts (Searle, 1997).

The inner city freeways may have been 'abandoned' but the DMR, through its powerful representation on the government's Urban Transport Advisory Committee (URTAC), introduced and employed a new rhetorical device to defend its role. 'Balance' was urged in future transport strategies, meaning some concession to community and environmental lobby claims against road building, without conceding, however, the 'need' for a freeway network. Governments embroiled in socio-political conflict around a particular policy setting are easily seduced by talk of 'balance': the URTAC view prevailed and major freeway corridors, totalling 117 kilometres, plus a major bypass of the CBD, were retained in strategic planning documents. In subsequent years, sections of the radial system were constructed, a strategy that maintained constant pressure for 'fill-ins' and extensions to resolve the congestion problems to which every new project inevitably gave birth.

From the early 1980s, the financial scenario began to tighten as federal funds for the DMR were reduced in a context of significant budgetary pressure for the state government. Nevertheless the State Transport Study Group (STSG) in 1985 produced a plan that projected major additions to the freeway network. (The STSG had relied in part on a seconded DMR officer for advice on demand modelling.) No additional major public transport infrastructure was recommended for new high growth urban areas. Interestingly, while the STSG's modelling 'showed' that patronage on existing rail lines would decline by 14 to 28 per cent, the same analysis concluded that, 'In the inner and middle suburbs traffic volumes under all population/employment scenarios could be accommodated by the existing road system'

(STSG, 1985, cited in Searle, 1997, p. 11). In short, most of the planned freeways were not needed.

In the late 1980s and early 1990s, major fiscal barriers emerged in the path of the public road building machine. Successive State governments, urged on by the DMR, turned to the private sector to fund major transport infrastructure works, in particular the tunnel under Sydney Harbour. The tunnel project terms set the frame for the subsequent series of privately funded transport infrastructure works that progressed the DMR's freeway plans during the 1990s. The project's financial risks were largely borne by the state government, and were deeply embedded in a mesh of complex, partially undisclosed contract arrangements.[4]

Nonetheless, a pattern for future freeway expansion had been set; one well adapted to the financial stringencies of the 1990s. New tollway projects were constructed and administered by private agencies (hardly *freeways*, now known as 'motorways'). A series of piecemeal motorway sections were completed in the south-west (M5), the west (M4), north-west (M2) and finally the east (Eastern Distributor) during the 1990s. While in many ways the projects were opportunistic and defiant of planning logic, they helped to ink in some of the remaining sections of the old DMR radial freeway scheme (see Figure 12.1).

By the late 1990s, grassroots opposition to the tollway model had emerged (Diesendorf, 1999; Wainwright, 2000a,b). Recognizing the significance of the discursive politics that helped to frame and secure private tollway construction as 'economic goods', one oppositional force constituted itself under the rubric 'Truth About Motorways' (TAM). At issue were the secrecy of the tollway contracts and their conditions, the environmental impacts of the projects and the veracity of the demand projections made by public and private agencies in support of the new roads. TAM had began a series of discursive and legal engagements with the new 'tollway-industrial' complex (i.e. RTA, State Department of Transport, and private motorway companies) that had so effectively produced a new freeway landscape in Sydney during the 1990s. In 2000, TAM accused one tollway provider of producing 'grossly over-optimistic traffic forecasts' in its project prospectus, after it was revealed that traffic levels on the new Eastern Distributor were lagging more than 18 per cent behind project estimates (Wainwright, 2000b, p. 3).

At the same time, another action group, Residents Against Polluting Stacks (RAPS), had emerged to oppose the tunnels that formed parts of the Eastern Distributor and (planned) M5-East (Wainwright, 2000a). These tunnels were conceived ostensibly on environmental grounds – to obviate the need to resume houses and land for freeways – but relied on major pollution emission stacks for ventilation of motor vehicle exhaust fumes. RAPS vigorously contested the RTA's risk assessment of the stacks, and eventually mounted an unsuccessful legal challenge against the projects.

In the 1990s, the RTA and other elements of government were able to adapt to, and eventually co-opt for their own purposes, a set of discursive

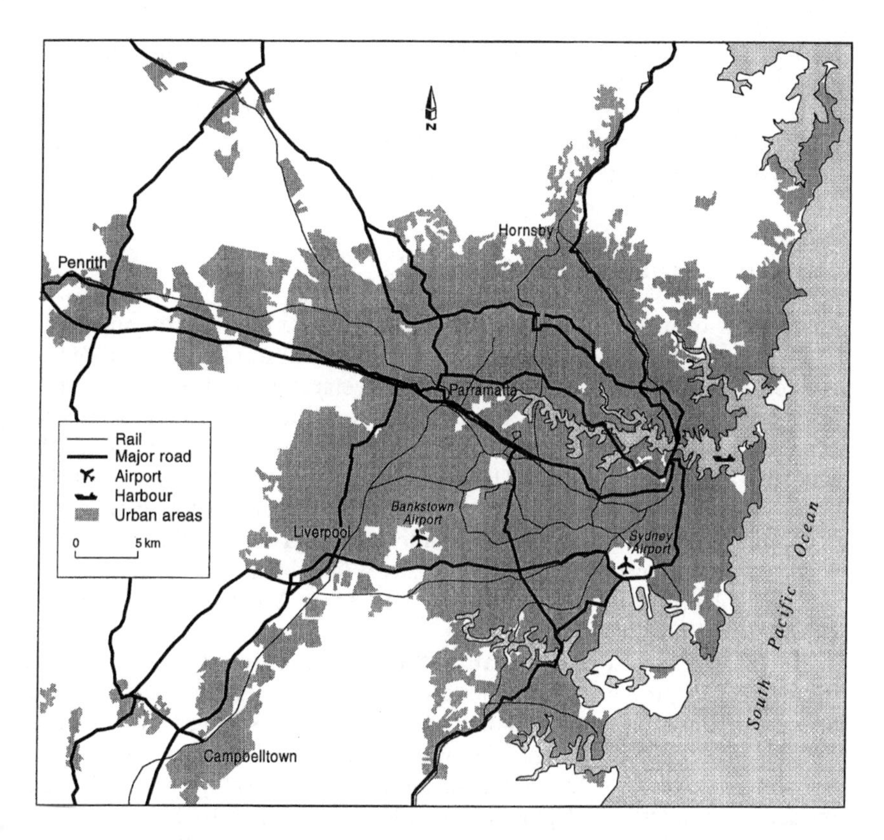

Figure 12.1 Sydney's transport system, 2002. (Map drawn by Chandra Jayasuriya.)

forces (notably neoliberalism) and structural economic changes (notably reduced public sector financing) that had undermined the basis for Fordist public infrastructure programs. The 'road machine' was also able to ignore increasing evidence of public dissatisfaction with state transport planning in general and with the lack of support for urban public transport in particular.[5]

The survival of Sydney's road program is a remarkable testament to institutional and professional resourcefulness in the face of profound dilemmas that were to overtake and obliterate other areas of public endeavour and their traditional aims (e.g. strategic social, economic and urban planning). It was also fortunate for the road builders that tollway projects emerged as relatively attractive objects for private financiers, against alternative investments, including other parts of the built environment (e.g. commercial and retail development). And of course, roads have always had their powerful patrons in Australian government in the form of Ministers for Roads/Transport who have traditionally enjoyed a far more powerful Cabinet status than their equivalents with responsibility for environmental and planning portfolios.

Finally, as Searle (1997) reminds us, we must not discount the robustness, not to say the tenacity, of the institutional cultures that have defined road agencies such as the RTA, where the engineering and construction of freeways has long been regarded as a sacred duty, or perhaps more correctly, the only true course of urban modernization.

Melbourne, 'On the Move!'[6]

The system of urban governance which prevailed in Melbourne up to the 1980s was one in which the task of building and running infrastructure was divided among a number of single purpose statutory authorities. Separate authorities handled tramways, trains, traffic and roads. Two others dealt with land use planning, the Melbourne Metropolitan Board of Works (MMBW) and the Town and Country Planning Board. By the 1950s an extensive system of suburban railways had made possible a spacious low density city organized along radial corridors of movement. The MMBW from 1956 acquired control over the city's main roads as 'one of the principal functions of town planning' (Anderson, 1994: 191 citing Victorian Parliamentary Debates 21, 1956: 2456).

The MMBW was constituted at the time by delegates of Melbourne's 52 municipalities under a permanent chair. In its Metropolitan Planning Scheme of 1954 the Board's planners set out a system of arterial roads based on the current radial patterns of movement determined largely by the land use patterns created by the radial railways. American examples provided the model for 'modern road practice': divided highway, restricted access and grade separated interchanges (MMBW, 1954: 90, 94, 100). The Board produced its highway plan in 1957 which included an inner city ring road, freeways to the east, south-east and north-west and a number of by-passes (Dingle and Rasmussen, 1991: 243).

In the 1960s, with the active intervention of the state Premier, two radial freeways were built. Then came another in 1971. The discourse at this stage referred to relief of congestion, the by-passing of 'bottlenecks' and 'oiling Melbourne's traffic machine' (Dingle and Rasmussen, 1991: 243). Even in this period there was much public protest, local government disquiet, and press campaigns against freeway construction (ibid.: 251). The inner ring road in particular was the target of protest.

The various separate government agencies with an interest in transport planning were brought together by the Minister for Local Government in 1963 as the Metropolitan Transportation Committee (MTC) which commissioned a transportation study by the American consultancy Wilbur Smith and Associates working with a team of local engineers – once again an example of the diffusion of American discourse. Mees (2000: 54) comments, 'The orthodoxy that population density determines the amount of travel by car had its genesis in the Chicago Plan, as did the dogma that improving roads does not add to traffic volumes'.

The report of the study published in 1969, closely following the lines of the Chicago Area Transportation Study, recommended a grid of freeways, at between 6 and 8 kilometre intervals, to meet Melbourne's transport needs up to 1985 (Figure 12.1). In contrast to Sydney, the tramways were, albeit unenthusiastically, retained, largely thanks to the forceful chairman of the tramways board (see Mees, 1999: 274). A new rail line was recommended to fill a gap in the radial system from the city core to the suburb of East Doncaster, together with an inner city rail loop which would permit the underground circulation of trains in the CBD (later built).

This freeway plan was adopted in full by the MMBW in its 1971 report 'Planning Policies for the Melbourne Metropolitan Region' which was then opened to comment. The reaction was rapid and furious. Inner suburbs, parts of which were once designated 'slums', were increasingly attracting young energetic gentrifiers who now found that the quality of their environments was to be destroyed by the construction of vast new roads and flyovers. From 1969 local residents campaigned to stop the Eastern Freeway through the Yarra river valley (Barricade, 1978) – but to no avail. Residents' campaigns with 'public meetings attracting thousands of people', however, did force the state Premier (Henry Bolte) to eliminate the eastern leg of the inner ring road (Rundell, 1985: 13).

In the face of intense public pressure the MMBW was forced by its constituent councils to publish a major discussion paper containing the objections to the scheme. Facing an election in 1973 a new Liberal Party state Premier announced that the proposed freeway network would be halved and the most controversial inner urban freeways dropped: 'freeways would be banished to outer suburban and country areas and greater emphasis placed on public transport' (Anderson, 1994: 244). Responsibility for metropolitan main roads was transferred to another agency, the Country Roads Board (CRB), which however, continued to make detailed plans for freeways.

A new battle erupted in 1975 against a plan to link the Eastern Freeway with the Hume Freeway to the north (F2). The attitude of the chief engineer of the Country Roads Board was revealed in a leaked confidential memo which stated that the Board's view was that the freeway would be built 'and therefore as far as the proposed study is concerned government objectives do not really matter' (Country Roads Board, 1975: 2). In 1980 a new transport minister ordered an inquiry into all aspects of the state's transport system and appointed a retired executive of General Motors (Australia), W.M. Lonie, to conduct it. The report of the inquiry, though referred to as the Lonie Report, was in fact written by the head of the Country Roads Board. It recommended sweeping cuts to public transport, including the new rail line proposed in the 1969 plan, and the revival of most of the freeways cancelled in 1973.

The action of cutting the inner heart out of the freeway network and leaving outer elements intact inadvertently gave the engineers a potent weapon in a radially organized city. Congestion inevitably built up in the gaps between freeways, creating demand for relief. Mees reports that although the Lonie report contributed to the defeat of the Liberal government in 1982, the new Labor government 'was gradually worn down by the persistence of the bureaucratic road lobby, rising road traffic and public transport deficits' (Mees, 1999: 148). Against continual local opposition Labor constructed new links between freeways (renaming them 'arterials'). Labor's metropolitan planning policy of 1987 incorporated, without additional analysis, the main radial and outer urban freeway routes from the 1969 plan. The plan also included an additional element which can be traced back to a plan produced by the Town and Country Planning Board in 1967 in opposition to the MMBW plan. This element was an outer orbital freeway linking a growth area to the north with one to the south and east of the metropolitan area (Town and Country Planning Board, 1967: 17).[7]

As part of its commitment to filling in some of the outer freeways proposed in the 1969 plan, the Labour government approved a circumferential freeway through Melbourne's western industrial belt and an extension of the Eastern Freeway (without taking up the proposal to build a new eastern suburbs rail line). In 1992, under the same fiscal pressure as New South Wales, the Labor government called for tenders for construction of a privately financed major expansion of the radial freeway system (including the western portion of the 1954 inner ring road) which was named by the succeeding Liberal government 'City Link'.

The pattern of freeway building pursued by the Liberal government (from 1992) and its Minister of Infrastructure – the same man who had commissioned the Lonie report – had thus all been prepared under the preceding Labor administration. The Liberals for the first time put planning and transport under the same minister in a single department. But the road engineers were placed in charge of strategic planning. The freeway plan of 1969, the engineers' 'holy grail' of half a century earlier, continued to exert its influence, as it still does. But the pattern of freeways remains radial following lines of peak movement demand which have not changed much since the 1950s.

The City Link private freeway development, in particular, became associated with the Liberal government's high risk policy style, and the unpopularity of placing a toll on existing roads ultimately led to the defeat of the Liberal party in the 1999 election. City Link doubled road space on about 16 kms of existing freeway, constructed a bypass around the CBD and two tunnels under parkland and the Yarra River, with exhaust towers rising in residential areas. The project was the largest BOOT (build-own-operate-transfer) scheme in Australia. The Victorian Government in the contract with the private operator also agreed to assist the channelling of traffic on to the freeways by closing certain public roads, limiting the capacity of others and not improving the public transport system in such a way as to divert transport users away from the City Link roads.

Under the 1990s Liberal government the western section of the ring road was completed, City Link was implemented, the extension of the Eastern Freeway was constructed, and plans were advanced to build the eastern section of the ring road. Only the surprise loss by the Liberal party of the 1999 election temporarily halted construction of the ring road. That halt, however, was shortlived and the Labor government, surprised by power and without a strong policy commitment of its own, reversed the decision and is moving to build a series of new freeways: a further extension of the Eastern Freeway, (tunnelling under part of the environmentally sensitive and beautiful Mullum Mullum creek), the eastern (Scoresby) section of the ring road and a link between the ring road and the Hume highway to Sydney. The State government's Infrastructure Planning Council after making some encouraging statements about sustainability, demanding that the 'true' social, environmental and economic costs of transport be recognised, talking about integrating the public transport system and changing behavioural patterns, then concludes that by 2020 'The road system will be fully connected [it was never unconnected!], including the completion of the metropolitan freeway and ring road system and other key regional [road] projects' (Victoria, Infrastructure Planning Council, 2001). In other words, whatever has to change it will not be the policy of building more and bigger roads.

The counterpart of the strength of purpose of the road engineers in Melbourne was the organizational inadequacy and managerial incompetence within the public transport system. Mees (1999: 149) has argued that the real weakness of public transport is the 'poor service quality and lack of integration, arising from an overall lack of planning'. As Mees points out, the problem was recognized by the MMBW in the 1950s but the post-colonial structure of governance was one of functional fragmentation. So 'it was not until 1983 that a single authority was established to run public transport, and that body never even attempted to integrate the different modes of transport' (ibid.: 151). There was no question of integrating the purposes of public transport with those of roads – inquiring, for example, whether an improvement of public transport service might help relieve road congestion and improve connectivity, or of comparing the costs and benefits of new public transport investments with those of new road investments.

Under these circumstances political influence on transport policy since 1973 has been feeble. The assessment of the chief engineer of the Country Roads Board in 1975 that the roads would be built irrespective of government policy can in retrospect be seen as not cynical but simply accurate.

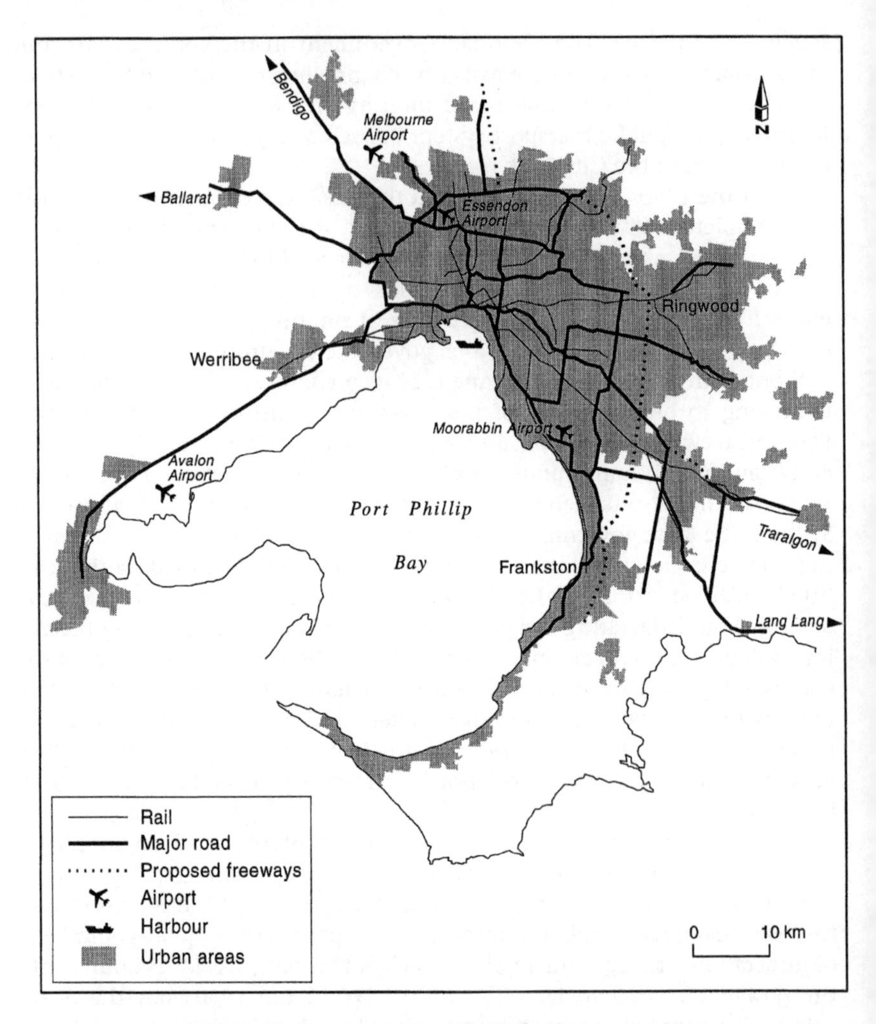

Figure 12.2 Melbourne's transport system, 2001. (Map drawn by Chandra Jayasuriya.)
Source: 'Challenge Melbourne; Department of Infrastructure, 2001.

While the spin different politicians have put on the policy has varied, politicians have followed wherever the road engineers have directed. The most striking aspect of the discourse of road building in Melbourne is the almost complete absence in official documents of a sense of debate – or even a recognition that there is or even could be a debate. This is not an absence of discourse but a discourse that renders opposing voices, such as that of the Public Transport Users' Association and the many community organizations opposed to road building completely silent, their presence invisible.

The only two forms of justification are the simple 'predict and provide' arguments unchanged since the 1969 plan: 'To provide a transport service matched to the expected demand throughout the design area' (MTC, 1969, Vol 3: 3), and the 'black box' of arcane cost benefit analysis and modelling whose assumptions are never tested and only rarely revealed. The environment, where it is mentioned, is treated as a matter of local pollution whose effects can readily be quantified in dollar costs – with ample scope for manipulation. Figure 12.2 shows Melbourne's transport system today.

Perth, 'A City for Cars'?[8]

There are similarities between Perth's transport story and those of its eastern sister cities. This is not surprising given that each city developed in much the same way – as a single coastal settlement supported by a huge resource-rich rural hinterland. There is, however, one major difference. Perth is geographically isolated and consequently rather introspective politically and culturally. Perth does not sit within an interconnected city region and has a smaller population than Sydney and Melbourne. Perhaps as a result of these factors there are some key differences alongside similarities in Perth's story compared with the other cities.

There have been three main phases of strategic planning, each of which makes a significant contribution to transport planning in Perth (Curtis, 1998). The Master Plan of 1955 (known as the Stephenson-Hepburn Plan after its principal authors) was essentially concerned with the physical use of land to accommodate population growth for a 50-year period. The plan attempted to integrate land use and transport planning. It was based on the notion of urban containment with high gross densities (22 dwellings per hectare), a single city centre and a secondary centre at Fremantle port. Transport was to include an integrated system of bus, rail and a network of highways linking centres and decentralized industrial estates (Yiftachel and Kenworthy, 1992).

This master plan formed the basis of Perth's first statutory plan the Metropolitan Region Scheme of 1963 (MRS). But by this time the city had begun to develop away from the 1955 plan. Low density urban development ensued at seven dwellings per hectare rather than the 22 per hectare envisaged by the Stephenson-Hepburn plan. There also was considerable growth in car ownership. The reaction to this was a dramatic shift in planning. Perth moved away from the compact city plan with its mix of transport services towards a 'demand led' approach. This saw the continuation of low density suburbia and a shift in transport emphasis towards the private car. This was facilitated through the provision of major freeways and expressways at the expense of planned railway lines to the north and north-east of the city.

The second phase of strategic planning was the development of the Corridor Plan of 1970 (Metropolitan Regional Planning Authority, 1970).

This continued the approach of the MRS, and so moved even further from the original compact city plan of 1955. The city was to be decentralized along four growth corridors with new sub-regional centres offering an alternative for employment and retail journeys to the city centre. Each corridor included substantial freeway, expressway and major arterial road provision. This approach was a result of a comprehensive regional transport study (PERT) (Perth Regional Transport Study Steering Committee,1971).

Like many planning studies conducted during the 1970s in the UK and America, the PERT study utilized a major land use-transport modelling exercise. As with Melbourne, the dissemination of American transport planning ideas was evident. Vorhees and Associates from Virginia, and Harris, Lange-Vorhees from Melbourne were employed on the PERT study. They followed in the footsteps of Americans De Leuw, Cather and Co who worked in Western Australia during the 1960s. The core objective was to predict and provide: 'continuously update the statutory plan so that it reflects the current and future demand for travel in the Region' (Perth Regional Transport Study Steering Committee,1971: 1–20). The study forecast less need for freeways than proposed in the 1963 MRS, yet recommended that none of the reserves in the MRS should be downgraded 'in order to retain flexibility for planning in the period beyond 1989' (Perth Regional Transport Study Steering Committee, 1971: I–11). This marked the point at which there was a major oversupply of road infrastructure, a situation still evident today.

The third phase in the strategic planning of Perth was 'Metroplan' (Department of Planning and Urban Development, 1990) (see Figure 12.3). This continued the outward spread of the four development corridors. A new corridor was added radiating north-east from the city centre. The plan provided for a continuation of radial road networks along each corridor to the Perth CBD (see Figure 12.3). It has been argued that this Scheme has been very effective at reserving land for roads, but less successful in delivering land uses in the right mix and location (Yiftachel and Kenworthy, 1992).

Like Sydney, the 'balanced transport' mantra of the late 1980s and through the 1990s has been successfully subverted by the road planners.[9] A notable example was the justification for duplicating the Narrows Bridge, a major river crossing for north–south traffic travelling to and around Perth city centre. This saw the doubling of road traffic lanes, from seven to twelve. 'Balance' was provided by the addition of a second dedicated bus lane. In another example, publicity by the state road building agency – Main Roads Western Australia (MRWA) – suggested that the addition of extra traffic lanes on the freeway from the city to the northern suburbs would improve the environment by reducing emissions (by reducing congestion). Always ignored is the obvious and well established fact that more roads lead to increased road traffic.

In 1962 all private bus services were brought under the control of the Metropolitan Transport Trust, the predecessor of today's 'Transperth'. While

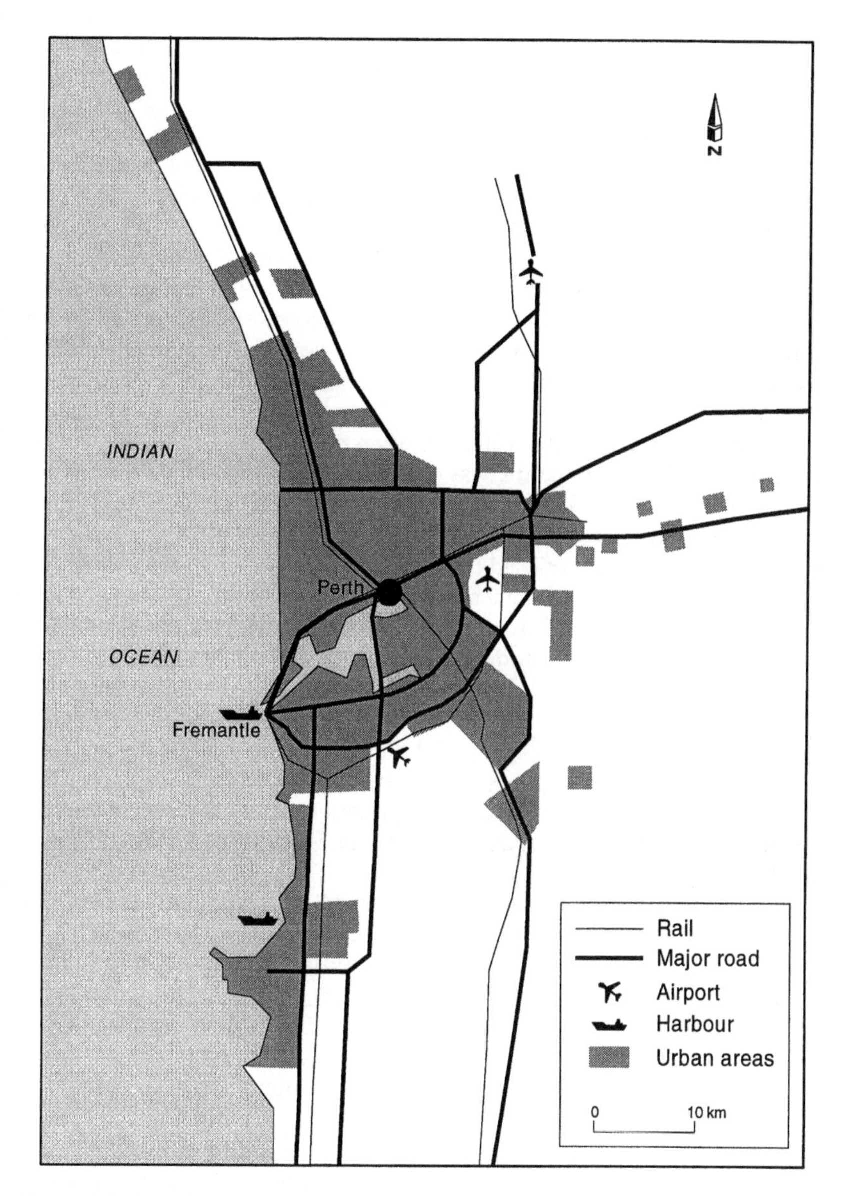

Figure 12.3 Perth's metropolitan strategy: Metroplan, 1990. (Map redrawn by Chandra Jayasuriya from original.)

Source: WAPC (2001) Future Perth Working Paper No. 1 – Metropolitan Perth – Planning Context.

Western Australia later followed the international trend towards privatization of bus services in the 1980s, the State government continued to retain control of service delivery, ensuring that community service obligations are fulfilled. In recent years a strong push to deliver improved bus services has been evident. The Ten-Year Plan for Transperth (Department of Transport, 1998) sets out to deliver one aspect of the 'balanced transport' strategy. It proposes to deliver a network of new limited express bus services with priority on congested links to connect centres of activity. The most successful example of this has been the introduction of a an orbital service called the Circle Route linking Perth's middle suburbs, major hospitals, tertiary education campuses, railway stations and park and ride car parks (Curtis, 2001: 2). Patronage was well in excess of forecasts almost immediately after the service began.

The rail system dominated early suburban development in Perth from the time the first line to Fremantle was opened in 1881 until the Second World War. After this new suburbs were developed away from rail lines, investment in rail-based transport diminished and patronage fell. By 1979, the Fremantle to Perth railway line was closed to make way for a new freeway – which has not been built. Four years later a new Labor government brought in a change of emphasis from bus-based to rail-based public transport. The Fremantle line was reopened in 1983 and public investment provided for the electrification of all three existing rail lines. A debate ensued about the delivery of a public transport service to connect the growing northern suburbs to the city centre. Transit planners proposed a bus-way that would run along the freeway median. Instead an 'expert review panel' recommended an electrified rail line and this option was strongly supported by the public. In 1988 the northern suburbs rail option was announced and work was completed in 1992.

In 1994, the route for the south-west (Perth to Mandurah) railway was reserved in the MRS (under a coalition Liberal/National government). In order to compete with the car, transit planners determined that the rail route must run down the freeway median strip. But locating the station in the centre of the freeway works against the creation of a walkable transit-oriented development. On the northern suburbs rail line, 60 per cent of users arrive at rail stations by car. Even at stations where bus interchange is provided, car access dominates.

Since the publication of the MTS in 1995, the promotion of 'balanced transport' has seen some recent public expenditure on alternative transport modes and a program designed to change people's travel behaviour called 'TravelSmart'. In addition to public transport, a state budget allocation of AUS$25 million over four years was provided in 1996 to deliver new and improved cycling infrastructure. This budget and program has recently been extended for a further four years but reduced from the proposed AUS$40 million to AUS$20 million.

Budget provision for different modes of transport has been overwhelmingly in favour of roads. Western Australia's Road Program 1998–2008 has a budget of AUS$ 5.1 billion (of which $1.3 billion is identified for spending on 'Transform WA' projects as a means of accelerating their implementation), but only AUS$ 84 million (1.6 per cent of the budget) is allocated to specific public transport projects (MRWA and DoT, undated). The delivery of a new railway line to connect the south-western corridor to the city centre is budgeted to cost AUS$1.2 billion. While provision of a new railway line is an impressive initiative, the priority of freeway building remains unquestioned and funds continue to flow for that purpose. Indeed, the budget for the new railway has only been made possible through the sale of a public asset – Alinta Gas. Private financing of freeways has not been evident in Perth. One argument has been that the effective planning of freeways through the MRS process has been so efficient that congestion has not occurred. Therefore road tolls would be difficult to impose as car travellers would have a myriad of other route choices. There is also fear of an electoral backlash from road users, the consideration being that they have paid enough through 'fuel tax' and registration fees.

Until recently there were three separate planning and transport agencies: the Transport Department (including public transport, cycling, walking and TravelSmart), Planning Ministry and MRWA. MRWA reported to one Minister, the other two agencies reported to a different Minister. Unlike the Transport Department, which could only utilise the operating budget for public transport, MRWA was able to access a separate capital investment budget. A recent (2001/2002) reorganization of state agencies created a newly 'integrated' planning and transport portfolio under a Department of Planning and Infrastructure (DPI). In this structure, the MRWA continues to exist, albeit with a service delivery function only. The road network planning function of MRWA has been located within the new DPI, but again as a clearly identified sub-department. By contrast, BikeWest, charged with promoting cycling and delivering infrastructure across the planning and transport agencies, has been quietly dissolved through reorganization.

For the most part, the public voice appears almost absent from professional and political discourse on freeway building. The proposals to build an inner city freeway were met with some opposition, but the debate focused mainly on whether to build at surface level or to tunnel, rather than on the need for the freeway itself. There is strong evidence from community surveys that road building is not supported by the public (Curtis, 2001) but this quiet counterdiscourse has been largely ignored.

In recent years there has been no shortage of State government reports (on air quality, congestion, car dependence and sustainability) that bring into question the continued provision of roads for private passenger travel (Curtis, 2001). However, there is little evidence that Perth's bureaucratic road building machine is seriously threatened. Outside government, the road building lobby continues to exert its influence. In 1993 a Western Australian lobby group called 'Fix Australia, Fix the Roads' was established. Its aims were to lobby for a greater proportion of the Commonwealth fuel excise to be returned directly to road building. The lobby group comprised, and was funded by, the Western Australia Royal Automobile Club, the Farmers Federation, the Transport Workers Union, Pastoralists and Graziers, and the MRWA, Department of Transport and local government. The group was renamed the Transport Foundation in 2000 and continues its advocacy of road building (Transport Foundation, 2000).

In transport policy discussion fora the economic case is now made that continued road building is needed in order to maintain the efficient movement of freight. There is little discussion about sustainable alternative solutions. For example, missing from the debate is any suggestion that 'freight only lanes' could be reserved on existing roads. Until 2001 there was scant discussion about moving road freight to rail.

In the post-war era, successive metropolitan planning strategies for Perth have been based on the notion of urban containment and integration of land use and transport planning. However, since the 1960s Perth has actually developed as a low density car-dependent city. MRWA has delivered a network of high capacity roads in advance of need and the resulting oversupply of road space has ensured that there is little suppressed demand. Critical attention has been diverted from the road building machine by an ongoing debate over public transport options (bus versus rail). Public transport planning itself has largely focused on park-and-ride systems where cars are necessary to access transit stations. Not surprisingly, public transport patronage and cost recovery are low. A more balanced and sustainable approach to transport planning has been promoted since the mid-1990s via the Metropolitan Transport Strategy. Subsequent initiatives have slowed the decline in the use of public transport and non-motorised modes. However, at the same time, there has been an acceleration in the road building program through the 'Transform WA' campaign. Again, the mantra of 'balance' has been used to deflect popular and scientific pressure for ecosocialization.

Conclusions

Taking account of the three metropolitan case studies surveyed above, the barriers to ecosocialization in Australia appear to be fourfold. First there is the absence of action at federal level to change funding priorities. Federal governments have continued to pour funds into roads without examining their effect on ecological sustainability. At state level the way that road space is paid for (in block payments for vehicle registration and exchange) compared with the way that public transport is paid for (in fares at point of sale) introduces an incentive to use the car wherever the choice between the two is available.

Secondly, the institutional divisions between road building and public transport agencies, the managerial weakness of the latter, and the absence of systemic planning, have combined to prevent the emergence of integrated solutions to transport problems. Thirdly, the discourses in formal policy documents and procedures supportive of road building have excluded discussion of alternatives. Congestion has been consistently presented as if it were an externally inflicted malady of the city that could be overcome through corrective surgery, sometimes radical in scale and in consequence. Further, it has been asserted that such cures, by clearing the sclerosis of urban arteries, would improve the circulatory capacities and therefore the productiveness of cities. Of course, no one among the road lobby ever seemed to ask whether the patient would survive the treatment!

While 'balance' has been adopted as a rhetoric, in practice the planning of roads and public transport as complementary systems has never been seriously contemplated, let alone implemented. In the three studies we have undertaken above, the 'balance' rubric seems to have been enlisted at various moments in response to surges in community and professional criticism of road building and car dependency. It is hard to argue against 'balance', which is a mainstay of conventional wisdom. If carefully deployed in policy strategies, the 'balanced approach' to transport planning seems to have a singular power to defuse and deflect critiques of the 'road machine'. Finally, in all three case study contexts, state politicians concerned with transport have not provided the leadership which could break the pattern of piecemeal road development to break open 'bottlenecks' and provide 'bypasses', thinking which has changed little since the 1930s.

Underlying all these barriers to ecosocialization is the resilience of the embedded alignment between the forces of social protection and economic liberalism. The political forces behind social protection have on many occasions become aligned with those behind environmental conservation but within the narrow framework of local conservation which is open to the charge of 'special interest' (NIMBYism) by the liberals claiming to be keepers of 'the public interest'. Australia's extensive, low-density cities are separated from each other by vast expanses of rural land. In this environmental context there has been little (obvious) evidence of the ultimate consequences of unrestrained road building: the destruction of the countryside, lethal concentrations of photochemical smog, and/or foreseeable terminal traffic congestion. In other national contexts, these same imperatives have manifested forcefully in recent decades and have thus stimulated new alignments of socio-political forces that have asserted the logic of ecosocialization. In Australia, complacency reins and will continue to do so until an inevitable and doubtless destructive intervention from nature produces a new socio-political outlook. We do not wish for a future enlightened by cataclysm: already, from here, it all seems so unnecessary.

Notes

1. See Beder's (2000, ch.15) expose on the Sydney Olympiad for a critical exploration of this example of the 'clean and green' myth.
2. While most of the rich countries agreed to reduce greenhouse emissions at the Kyoto summit, Australia was granted the right to increase its emissions up to 2010 by 8 per cent on 1990 levels. Current forecasts by the Australian Bureau of Agricultural and Resource Economics estimate that the increase will be nearer 40 per cent (Davidson, 1999: 17).
3. In 1901, six self-governing British colonies united within a federal structure to form the Australian nation. Today, the federal system comprises six states and two self-governing 'territories'. In this chapter, for the sake of simplicity, we will use the term 'states' to refer to the second tier of Australian government, which also includes the two territories.
4. 'Commercially ordained' privacy was a key feature of the tunnel and subsequent freeway projects. In 1994, the State Auditor-General, after observing that the benefits and risks of the tunnel project were largely borne by the Roads and Traffic Authority (RTA) (successor to the DMR), concluded that the venture was public not private. The tunnel also failed most benefit-cost ratio tests.
5. The most recent example of this evidence was a survey undertaken by the Warren Centre of the University of Sydney which showed 'strong support for allocating some road funds to public transport' (Dobinson, 2002: 15).
6. 'Victoria On The Move' was adopted the slogan by the State government of Victoria in the 1990s and appears on all car number plates issued during that period. Allegedly the first government to have 'On The Move' as a slogan was that of Mussolini in pre-war Italy.
7. Though the 1929 Report of the Metropolitan Town Planning Commission included a 'Circumferential Route' as part of a network of 'intersuburban roads' (Metropolitan Town Planning Commission, 1929: 92 and passim).
8. This sub-title is taken from a car sticker slogan devised by a local sustainable transport advocate. It was an ironic response to the City of Perth's CBD marketing campaign which promotes 'Perth – a City for People'. However, the City relies heavily on revenue from the very large number of public parking spaces it supplies – another publicity campaign trumpets that 'Your car is as welcome in the City as you are'!
9. Balanced Transport' is the main objective found in the Department of Transport's 1995 Metropolitan Transport Strategy (Department of Transport, 1995).

Part III

Best Practice in Sustainable Transport

13

Towards Sustainable Urban Transportation in the European Union?

Rolf Lidskog, Ingemar Elander and Pia Brundin

Introduction

Issues related to transport are global in the sense that they are possible to find wherever a city of certain size is located. The transport sector is a major source of CO_2 emissions, with urban transportation generating more than half of these. This means that any response to the climate change issue and any efforts to meet the Kyoto commitment must include changes within the urban transport sector.

However, urban transportation also generates environmental problems that are of a more local character – traffic congestion, low air quality and health effects, to mention just a few. In this case there is no such strong interdependency as in the case of the threats facing the global commons. Thus, a constructive and far-reaching proposal for the transformation of the transport system *within* a city could be implemented without co-ordination at the international level. Towns and cities are also in a position to develop new urban transport systems relatively independently from central government decisions, although major investments cannot be made without monetary support from national or international bodies. This means that the opportunity structure for the creation of new local transport systems seems to be more open compared with the case of the global commons. On the other hand, people's dependence on the car is a multifaceted phenomenon strongly related to modern culture and life style. Thus, changing people's transport behaviour is not a simple task.

Transport has an overriding role in European economic activity and growth. However, although massive investments in public transit have recently been undertaken in European cities – largely within the framework of the Trans-European Networks (TENs) initiative – just-in-time delivery and other new logistic concepts together with increased used of private cars for travel to work, shopping and various leisure activities are obvious threats to the development of *sustainable* transport.

Indeed, urban transport in its present form is far from sustainable, and – what is worse – the overall development goes in the wrong direction. Today, the transport sector is to blame for 28 per cent of the CO_2 emissions in the European Union (EU). If nothing is done at the EU level to reverse the trend, the emissions from transport will increase by 50 per cent, from 739 million tonnes in 1990 to 1,113 million tonnes in 2010 (EC, 2001: 27). Nevertheless, at all levels of government in Europe a higher level of priority now seems to be given to building and maintaining a fast, comfortable and reliable system of public transport, and in many European cities there is a growing concern about the problems connected with the continued growth of numbers of automobiles. The indicators of this trend vary from city to city, but include the development of rail, metro, bus, tram or combinations of these transport systems. Extensions of existing heavy rail systems (the Paris Metro), new heavy rail systems (Brussels, Amsterdam, and Vienna), new express rail systems (the S-bahn trains in many German cities) and high-speed inter-city rail are some examples (Hall, 1995). In addition, projects to stimulate walking and cycling are abundant. The general idea behind these projects is that giving higher priority to cyclists and pedestrians in traffic planning will make more citizens leave their cars at home.

The aim of this chapter is to draw attention to the various efforts made by European cities to reduce their dependency on cars for transport and mobility. Although transport clearly interacts with cities at both long-distance and intra-regional levels, the focus in this chapter is on transport *within* cities. The chapter is organized in four main parts. The second part briefly surveys the problems associated with urban transportation and the possible solutions discussed in the policy literature. The statements made at the EU level since 1992 with the Fifth Environmental Action Programme and the Green Paper is the topic of the third part, whereas the fourth part reviews the initiatives taken by networks of cities, and analyses the different types of projects aiming at the development of sustainable urban transport. The chapter concludes with an assessment of the strengths and weaknesses of the efforts so far undertaken by European cities to diminish their dependency on the motor car, a creation of modernity famously described by Roland Barthes as 'the exact equivalent of the great Gothic cathedrals' (Barthes, 1972: 88 quoted in Sheller and Urry, 2000: 737).

Urban transport: problems and solutions

Obviously the urban transport system is dependent on a number of contextual factors such as the form of the city, the employment structure of the region, the level of welfare of the nation, and people's life styles. However, at the same time cities all over the world seem to face the same kind of problems, even if the size varies considerably. In Europe 79 per cent of the population live in urban areas. In the 1990s urbanization increased, and the

Table 13.1 Per cent modal share of person movement in Western European cities, 1870–1990

Mode	1870	1930	1990
Walk/cycle	91	29	10
Other private	4	10	71
Public transit	5	61	19

Source: Adapted from Hart (2001), p. 108.

larger cities continued to grow, although there was a migration wave from the city centres to suburbs or nearby towns. Business to some extent shifted from central locations to outlying areas (Mullally, 1997: 287). This means a growing need for mobility, and to a large degree this has resulted in eco-detrimental forms of transport, i.e. more cars in the streets. External location of work places, supermarkets, amusement parks and other amenities implies a growing need for transportation. Thus, we face a vicious circle: car transport makes urban sprawl possible and urban sprawl makes suburbanites auto-dependent (Freund and Martin, 1993: 20). Measuring the share of personal movement in cities in Western Europe shows the overriding dominance of the car (Table 13.1).

The transport sector has a wide range of direct and indirect impacts on the environment. Road transports in particular are responsible for significant levels of anthropogenic greenhouse gases, carbon dioxide, methane, chlorofluorocarbons and low-level ozone precursors. This problem is related to the massive use of fossil fuels, and besides global warming the scarcity of this energy resource is one of the most important global environmental challenges (see Chapter 3 of this volume). Thus, the road transport sector consumes a large amount of finite resources, but also materials for construction of parking lots, service stations, etc. (see Chapter 7 of this volume).

Aside from emitting eco-damaging particles and extensively consuming unsustainable resources the road transport sector puts high demands on land and severely contributes to traffic congestion in cities. The land-use share of the transport system in modern cities is estimated to be 20–35 per cent of the total urban space (OECD, 1996: 20). Of course, this is reflected in ever-increasing motor traffic, with ensuing congestion. Thus, OECD statistics estimate that the average speed of vehicles in major European cities has declined by 10 per cent over the last 20 years. The average traffic speed at peak times is now even lower than in the days of horse-drawn carriages (EC, 1996a: 12). The cost of congestion is estimated at approximately 0.5 per cent of gross domestic product (IP/01/1263). However, it is hard to 'build one's way out of congestion' because new road construction generates more traffic (Freund and Martin, 1993: 20; Mogridge, 1997; Fuji and Kitamura, 2000; Black, 2001).

Obviously, the North American ideal of an urban landscape with a majority of the population accommodated in decentralized, auto-based, low density urban forms attracts both Europe and the rest of the world. However, this ideal also meets resistance by a European urban tradition putting a higher value on land use policies to reduce urban vehicle kilometres, and promoting higher shares of movement on foot and bicycle and by public transport. To counteract the tendency towards Americanization, the overriding ideology in European urban policy propagates densification – as implied by the catchword *the compact city* (see Chapter 2 of this volume).

The compact city ideal implies that concentration leads to lower fuel consumption than dispersal, and that the average density of an area is a key predictor of energy use. Empirical studies suggest that 'even the poorest households find they cannot do without a car in dispersed areas, while conversely in accessible inner urban areas some of the richest households choose not to own a vehicle (or not to invest in a second one)' (Barton, 2000: 110). Although data seem to point in the direction of a positive correlation between sustainable transport and density in urban areas, the compact city ideal could be challenged on other grounds, e.g. people's need for green space or children's need for playgrounds (cf. Breheny, 1992).

The emissions and the congestion produced by extensive car use severely affect people's health. Air pollution from road vehicles is a prime cause of rising levels of asthma and other respiratory illnesses, whereas traffic congestion contributes to stress and road accidents. Road traffic in the European Union yearly causes more than 40,000 deaths, 1.7 million injured, and a social cost estimated at 160 billion Euro[1] every year (IP/01/1263; RTD info, 2001, No. 32). Technological innovation (catalytic converters, bumpers and road barriers) may reduce these problems, although reducing the absolute level of vehicle use, improving the public transport system, and increasing pedestrian and bicycle accessibility would promote the health of urban citizens in a much broader sense (Barton, 2000: 54–5).

With no ambition to give a comprehensive view of the problems connected with urban transport, and possible solutions to solve the problems, the following table at least gives a rough summary of problems/solutions (Table 13.2).

Between market liberalization and sustainable mobility: towards a common European transport policy?

A common transport policy (CTP) for the European Union was earlier announced as an important issue, and has been the subject of many Commission proposals. So far, however, it has proved an elusive goal, although blueprints for such a policy have not been lacking. For a long time the major objective for CTP was to get rid of the formalities at the Member State borders to make transport flow smoothly between countries and

Table 13.2 Problems and solutions in urban transport

Area	Problems	Proposed solutions*
Emissions	Global warming	Ecocar
	City air quality (fine particles, etc.)	Public transport (ecobus, metro etc.)
	Land contamination	Park and ride
Energy use	Scarcity of finite resources	Bicycles and low impact technology
Land-use and urban form	Threat to green areas	Car-free cities
	Diminishing biodiversity	Telecommuting
Time/stress	Traffic congestion	Green corridors
	Lack of parking area	Densification
	Delays	Smart cards
	Unpredictability	Road pricing
Health/safety	Noise	
	Accidents	

Note
*N.B. The solutions presented in the table are not related to any specific area or problem.

ensure freedom of movement for persons and goods within the Common Market. Today a mixture of harmonization and liberalization has come to the fore as the twofold top priority: co-ordination of investments and common rules of the game would allow for lower logistical costs, while an integrated open market would deliver competitive benefits (McGowan, 2000: 460).

Although market liberalization has been the overriding objective of European transport policy, other elements began to gain a higher profile in the late 1980s. Thus the Fifth Environmental Action Programme in 1992 brought sustainable transport high on to the rhetorical agenda of the European Union (COM, 1992, No. 23: 'COM' denotes Commission of the European Union). The programme underlined the role of transport as one of the main causes of air pollution and noise nuisance, and presented a range of proposals for the integration of environmental aspects into transport policy. Subsequent Green Papers from the Commission emphasized the need to integrate environmental aspects into CTP (COM, 1992, No. 46; COM, 1992, No. 494; COM, 1998, No. 466). Suggesting a number of ways in which the use and development of public transport, bicycles and walking might be encouraged in urban areas, the stated objective of the Commission is to offer easily accessible and affordable alternatives to car use. The Commission advocates better access to public transport, better co-ordination of the different modes of transport and a greater focus on public transport in research and regional development.

The integration principle

The Amsterdam Treaty, signed in June 1997, stated that responsibility for the environment should be integrated into the design and execution of EU policies. Article 2 of the Treaty states that one of the basic Union objectives is to promote economic and social progress and a high level of employment, and to achieve balanced and sustainable development. The integration process started in June 1998 when heads of government of the Member States met in Cardiff. At this meeting, the European Council endorsed the principle that major policy proposals of the Commission should be accompanied by appraisal of their environmental impact. With regard to the transport sector the integration principle aims both at measures relating to the improvement of infrastructure and measures to control the CO_2 emissions. Published by the Commission in June 1998 the document *Developing the Citizens Network* (COM, 1998, No. 431) announced initiatives to measure the performance of transport systems and to assess the strengths and weaknesses of the transport system in cities and regions.

Sustainable mobility

The CTP was initiated as a way of providing for the free-flow of goods and services, labour and capital across the national frontiers between the Member States. A common policy of infrastructure for high speed trains, the liberalization of railways, ideas for integrated traffic and the internalization of external transport costs by new taxation formulas are main issues in the current European policy, which is mainly based on the action programme *Sustainable Mobility: Perspectives for the Future* (COM, 1998, No. 716). The aim of the CTP is to encourage the development of efficient, socially acceptable, safe and environmentally friendly transport systems. In particular, the efficiency of transport systems is regarded as a fundamental objective for the competitiveness of Europe and for growth and employment. Although targets for environmental quality, as well as reduction targets for the Union as a whole (i.e. the Kyoto commitments), have been adopted by the European Union, these are largely to be classified as 'soft laws' with weak or unclear legal status (cf. Bernitz and Kjellgren, 1999: 40). Thus, sector-specific targets are still rare at the EU-level, although most of the Member States have adopted objectives and targets for each country's own national transport sector.

The Sixth Environmental Action Programme

In January 2001, the Sixth Environmental Action Programme *Our Future, Our Choice* was presented by the Commission (COM, 2001, No. 31). The programme identifies four priority areas – (i) climate change, (ii) nature and biodiversity, (iii) environment and health, and (iv) natural resources and health – but has been criticized by environmentalists for its absence of clear targets and timetables for implementation. The programme underlines the importance of continued efforts to implement the Integration Principle

Table 13.3 Environmental impact of different modes of transport

Transport mode	Environmental impact	Measures taken by the EU
Road transport	*Acidification*	*Acidification strategy*
	Climate change caused by (among others) CO_2	Proposal to decrease CO_2 emissions from private cars
	Land-use	
	Smog	
	Noise	Community directives set maximum sound levels
Aviation	*Noise and gaseous emissions*	*Strategy to improve technical standards and related rules*
		Community directives set maximum sound levels
Shipping	*Acidification – pollution emissions of nitrogen (NO_x) and sulphur dioxide (SO_2)*	*International convention for the prevention of pollution from ships (MARPOL)*
Railway	*Land-use, noise*	Community directives set maximum sound levels

(mentioned above), but does not present any new strategies in the transport area. The Commission also states that particular attention should be given to aviation emissions, which are expected to grow by almost 100 per cent from 1990 to 2010. Further, the Commission underlines that the Member States as well as regional and local authorities are responsible for many of the steps that need to be taken in the transport policy area. In Table 13.3 the environmental impact of different modes of transport is summarized as well as the measures decided upon by the EU.

Thus, *l'acquis communautaire*, i.e. the comprehensive body of EU treaties, regulations, directives, decisions, recommendations etc. that sets rules for the Member States, includes a large number of Green and White papers, programmes and initiatives aiming at sustainable development in general and sustainable urban transport in particular. Other important means for stimulating sustainable urban transport in the EU are the tax policy and structural funds, as will be briefly described in the following pages.

Fiscal policy

The Member States of the European Union show significant differences in the structures and rates of taxation on fuels and vehicles across the EU. This could be explained by historical specificities and different political priorities on the part of the Member States. The planned enlargement of the EU by the addition of the countries of Central and Eastern Europe is expected to lead to a continued slow progress in the fiscal policy area for the years to come.

The Polluter Pays Principle and the requirement to internalize external costs are important principles in EU policy. A number of initiatives have been made in order to put the principles into practice. Progress to date is however rather limited. Fiscal harmonization has been the lowest common denominator, with some basic minimum rates set that have now been overtaken by fuel duty increases in most Member States. The lack of progress could be explained by the unwillingness of a number of Member States to hand over any fiscal competence to the Community. In addition, fiscal policy currently requires unanimity within the Council.

The Dutch Government in 1999 made proposals for an *Eco-Schengen*, thus allowing a particular EU state to proceed on its own if the wider negotiations proved to be unsuccessful. The sector integration strategies of the Council and the need for a more coherent climate strategy in the transport sector might present additional opportunities for further harmonization of environmental taxation. In its latest White paper on transport policy, the Commission proposes a wide range of measures to develop infrastructure charging which take into account the external costs and encourage the use of the least polluting modes of transport.

Structural development

The EU Structural Funds aim to help reduce economic and social disparities in the Union. Transport and infrastructure plays an important role in the projects supported by the Funds. The real start of involvement by the Union in infrastructure policy was in the beginning of the 1990s, with the Maastricht Treaty (signed in February 1992). Since then, the budget for regional policies has increased, today representing more than 35 per cent of the total budget for the EU. The European Treaties set out the general principles of sustainable development and environmental protection, while the Funds constitute some of the most important financial instruments to support the implementation of Community policies. The Funds are, therefore, potential policy instruments for implementing sustainable development. The environmental considerations in the Fund procedure to date are, however, rather limited. This could partly be explained by the mostly modest degree of influence exerted by the environmental authorities in the Member States (Whitelegg is particularly critical of EU policy; see Chapter 7 of this volume).

Since 1989, the Structural Fund and the Cohesion Fund have been major sources of finance in the development of Europe's transport infrastructure. Transport needs of the weaker regions of the Union are not the same as those of the stronger regions. A balance between the different modes in the different European regions is difficult to establish. The most important goal for Europe's transport and funding policy is the reduction of unfavourable environmental impacts of transport. The aim of the action programme Agenda 2000 is to strengthen the Community policies and to set a new financial framework for the period 2000–06 (COM, 1997, No. 2000). Transport is an

Table 13.4 EU directives and regulations with influence on transport infrastructure

Legislation	Subject
Regulation and decision over structural funds	Different regulations establishing the European Regional Development Fund and the Cohesion Fund.
Decision 98/2179/EC	Integration of environmental policies into other policies.
Council directive 85/337/EEC and directive 97/11/EC	Environmental assessment of the effects of private and public projects.
Council directive 90/313/EEC	Directive on the freedom of access to information on the environmental imposes a duty to ensure that information, which is normally held by publicly accountable bodies.
Council regulation 92/1973/EEC	A regulation that establishes the financial instruments for the environment known as LIFE (actions which support the polluter Pays and Subsidiary principles of the Community Policy). LIFE only funds a proportion of the cost of the project.
Council regulation 95/2236/EC and 99/1655/EC	The regulatives lay down general rules for the granting of Community financial aid in the field of TEN (for the periods 1995–99 and 2000–06).
Decision 96/1692/EC	Community guidelines for the development of TEN.
Council regulation 99/1655/EC	Extends the scope for intervention of TEN projects, particularly in order to stimulate growth in investments in the form of risk capital. Recourse to private sources of funding is expressly encouraged.
Decision 98/EC	Indirect influence on construction and reconstruction of infrastructure through noise and emission standards.
Council directive 79/409/EEC and 92/45/EC	Conservation of wild birds and conservation of natural habitats and of wild fauna and flora (parts of Nature 2000).
Council directive 2001/12, 2001/13, 2001/14	Rail infrastructure packages. Aims to improve the effectiveness of Trans European rail freight.

area of priority of the Agenda as well as the objective of integrating environmental considerations into the work of different Funds.

To summarize: although striving for a common transport policy (CTP) the European Union has produced a growing number of proposals, recommendations and directives in the past ten years, particularly in the era of market liberalization (see Table 13.4). But McGowan's point is valid:

> It remains debatable whether, taken together, the various EU initiatives constitute a CTP. The tensions between different programs – between the competition ethos of liberalisation and the planning ethos of the TENs, and between the growth-led perspective of the TENs and the sustainability of environmental policy – make it hard to reconcile the conflicting objectives inherent within EU transport policy (McGowan, 2000: 481–2).

Despite this conflict, however, the EU has a crucial role to play as facilitator of local initiatives and provider of economic resources for the development, testing and evaluating projects aiming at more sustainable urban transport. The next section will present a number of such initiatives, some proposed by the EU, others emanating from local authorities and NGOs, but in many cases supported by EU finance.

Pioneering cities: a way forward?

The task of creating sustainable urban mobility within the EU demands activities at all levels of policy. At the EU level, the European Commission has been invited by the Council to facilitate the exchange of information and to develop a comprehensive set of indicators of transport sustainability and tools for evaluating external costs, with the European Environmental Agency.[2] The Council has invited the Member States to draw up and implement national and local strategies with targets for reducing the level of road traffic growth and the environmental impacts of transport. The EU also facilitates different local initiatives and projects.

In its latest White paper for transport policy, the European Commission supports cities pioneering the development of urban transport, and sees the exchange of good practice as a central tool in creating a sustainable mobility within cities (IP/01/1263). Obviously, there are projects run by other actors than the EU. The Friends of the Earth organizes different campaigns against car-based development (www.foe.co.uk). Critical Mass organizes monthly bicycle rides with the aim of reclaiming the streets from the car (www.criticalmasshub.com). Together with the European Conference of Ministers of Transport, the OECD has initiated a project on urban travel and sustainable development where the task is to investigate and overcome barriers to implementation of integrated sustainable transport policies. Over the period 2000 to 2006 a total of 1 billion Euro will be made available annually in the field of environment and transport for the candidate countries in their preparation for accession to the EU, and this will probably also include efforts to create sustainable mobility in cities (www.inforegio.cec.eu.int). The following description of local initiatives, however, is restricted to projects initiated or supported by the Union.

Towards a new urban mobility culture

A number of European Commission initiatives for more sustainable transport have focused on research programmes and demonstration projects with zero and low-emission vehicles, such as Jupiter 2 and CENTAUR. Other projects that could be mentioned are CITELEC (a network of cities interested in electric vehicles), Alter (a network of cities who have declared their commitment to promoting cleaner vehicles in Europe) and POLIS (a network of European cities and regions, which are working together on transport and environmental problems via innovative transport technologies, policies and funding mechanisms). ZEUS is another project devoted to providing citizens with the opportunity to travel using low-energy modes of whatever form of transport they choose. The main task for the project is to help remove market obstacles which hinder the widespread use of zero and low-emission vehicles. ZEUS is one of several Targeted Transport Projects (TTP) supported by the THERMIE (a demonstration programme) at the European Commission. Member cities of the ZEUS project are Stockholm (Sweden), Helsinki (Finland), Copenhagen (Denmark), London and Coventry (UK), Bremen (Germany), Luxembourg (Luxembourg), Palermo (Italy) and Athens and Amaroussion (Greece).

CIVITAS (City-VITAlity-Sustainability) is a project aimed at supporting the best integrated and most innovative proposals for the development of urban transport by committed European cities. By encouraging competitive alternatives to the use of cars in city centres, the ambition is to prevent European cities from growing congestion and pollution. Pilot projects in Vienna, Stuttgart, Oxford and Nantes have shown innovative solutions, which are planned to be extended and built upon. Cities wishing to participate in CIVITAS are required to implement packages of integrated measures, possibly including access restrictions for polluting vehicles, charging for urban roads, encouraging new types of mobility and promoting clean and efficient urban public transport. In October 2001, under the Fifth Research and Development Programme, 50 million Euro were earmarked to support 14 pilot cities for integrated and innovative projects to combat congestion and pollution (IP/01/997).

The Car Free Cities (CFC) was launched in 1994 in the context of the Fifth Environmental Action Programme *Towards Sustainability*, as a joint initiative of the European Commission and the organisation Eurocities (www.eurocities.org). It gathers over 70 local authorities from 20 countries (both EU and non-EU). CFC includes a political commitment, The Copenhagen Declaration. By signing the declaration, the members have committed themselves to transforming the network into an operational mechanism for improving the quality of life in cities.

The main policy goal of CFC is the promotion of a new mobility culture in Europe. The most important activities are carried out at the local level, where the goal is to help local decisionmakers develop a mobility policy based on the reduction of the use of the private car and the promotion of more environmentally friendly modes of transport. The main task for CFC is to facilitate the exchange of knowledge and the development of projects, especially the identification and transfer of good practice and successful initiatives between cities.

A part in this work is the 'European Car Free Day'. Each year on 22 September European cities are called to make it possible for the public to choose transport means other than the car. It was launched in 1998 in France, and has developed to a Pan-European event in 2001 with more than 600 cities participating – including almost all capitals from the Member States. This annual event aims to encourage the use of alternative forms of transport and travel other than private cars, not least to enable city dwellers to discover other means of transport and to experience this day without restricting their mobility.

The European Car Free Day and other activities promoted by CFC are attempts to change behaviour and attitudes, as well as raising awareness of a new mobility culture. The promoters believe that it is only through raising awareness of the problems and their possible solutions that real progress can be made in cutting the environmental, economic and human costs associated with private car use.

A key instrument: knowledge exchange

The exchange of good practice has become one central means for the Commission in the development and implementation of a new approach to urban transport. ELTIS, The European Local Transport Information Service, is an Internet guide to current transport measures, policies and practices implemented in cities and regions across Europe (www.eltis.org). The aim is to support the practical transfer of knowledge, exchange and experience in the field of urban and regional transport in Europe. ELTIS is jointly funded by the transport research and policy directorates of the Directorate General of Transport of the European Commission and the International Union of Public Transport (UITP). ELTIS has also developed partnership with other actors, such as the above-mentioned POLIS. The Car Free Cities network contributes significantly to the ELTIS data base.

An Internet investigation of projects on environment and transport produced 198 hits on the ELTIS homepage (January 2001). Denmark, the Netherlands and Belgium have together contributed 87 of the 198 projects, and European capitals and big cities are best represented among the cities. The projects could be divided into categories as follows from the table below (Table 13.5).

The projects on walking and cycling range from play streets for children in cities to bicycle tracks and bridges, benches and rest poles for pedestrians, and priority for snow cleaning in cycling routes. Among the less usual projects is the bicycle lift in Trondheim, Norway, where the cyclist is pushed up

Table 13.5 Projects on environment and transport in ELTIS

Walking and cycling	Transport and land use	Public passenger transport	Traffic organization, urban planning	Electric vehicles	Other projects
84	14	17	63	13	6

the hill at a speed of seven kilometres per hour. Another example is the Floating Distribution Centre, run by DHL Worldwide Express, where boat transports on the canals of Amsterdam serve as the base centre for bicycle couriers, thereby decreasing the need of van transports in the city centre.

The traffic projects include traffic calming and speed-reducing measures, bus priority in cities, constrained car use (i.e. limited access to the city centre on Saturdays), school crossing guards and parking spaces for car-sharing vehicles. There are also several research programmes represented on studies of behaviour patterns and on environmentally friendly vehicles. Surprisingly few of the projects concern public passenger transport. Although this may partly be explained by deficits in data collection – projects on public transport not being reported to ELTIS – this is of interest because of the function of ELTIS as a transmitter of knowledge.

The projects focus on carrots rather than sticks. Educational and informational projects are most common, which could be explained by the need to break existing patterns of behaviour to attain a more sustainable approach to transport. The general idea behind most projects seems to be that higher priority for cyclists and pedestrians in traffic planning will make more citizens leave their car at home. Thus, the priority of sustainable mobility within cities is high in the EU, at least at the level of rhetoric. Reflecting a belief that this is the best way to change behaviour and attitudes, at the local level there seem to be many educational and informational projects. However, the emphasis on information and education could also be explained by the fact that this is a non-conflict choice of action. Handing over much responsibility for attaining a sustainable transport system to the local level and to the citizens is a 'cheap' strategy for the national and international levels of politics, and an easy way out of the conflict between the two diverging political targets of environmental sustainability and economic growth.

Furthermore, mobility is an essential part of the modern identity, and the choice of transport a crucial element in citizens' wider struggle for time (Jensen, 2001: 16). If responsibility for urban transport is placed on the citizens' own shoulders, it is important to create a situation where there are opportunities to develop viable transport alternatives. If not, there is a risk that the changes will have drastic consequences for the everyday life of the citizen, or that there will be no change at all.

Conclusions

Mobility has been one of the major goals of the European Community since its beginning. Together with agriculture, competition and external trade, transport was one of the first common EU policies. The goal was to make transport flow as smoothly as possible among countries, thereby ensuring the freedom of movement of persons and goods within the Common Market. In 1985 the Single European Market was launched, and in 1993 it was finally completed. The single market was not only based on harmonization – the creation of European rules and standards, thus creating a level 'playing field' – but also on liberalization: through transport earlier national markets were opened up for competition. An efficient transport system is therefore seen as a prerequisite for the Single European Market. As stated by the EU: 'In an increasingly global economy, our transport system must not be left behind' (CTP, 1999).

The point of departure for the common transport policy is that transports are 'the life blood of our economy', that a modern society and competitive economy are unthinkable without an efficient system of transport. With the concept 'sustainable mobility', the Union seeks to reconcile the need for mobility with the imperatives of safety, respect for the environment and social responsibility. Like the concept 'sustainable development' with its economic, ecological, social and cultural dimensions, the concept 'sustainable mobility' sees its dimensions as inseparable from one another. Thus, if any one of these dimensions is neglected there will be no sustainable mobility.

At the same time, the EU has declared that the volume of transport cannot be reduced, because the demand for mobility is ever-increasing and any prohibitive measures would seriously hamper the working of society. While the transport sector is a major source of CO_2 emissions within the EU, doubts may be raised whether a reorganization of the transport system through technological development and changing modes of transport is forceful enough to make it possible to reduce the CO_2 output in a situation where the volume of transports continues to increase (see Chapter 7 of this volume).[3]

Within the EU urban transportion generates about half of the transport sector's CO_2 emissions. A change towards more sustainable urban transport therefore serves not only to create healthier cities, but will also be a major element in combating climate change. From that perspective sustainable urban transport is of vital importance for the EU when approaching its Kyoto commitment. Urban transports are not only associated with the problem of climate change, but also with a wider collection of environmental problems. Current urban transports have local effects on human health, they cause accidents, they are a threat to green areas, they create traffic congestion and they impoverish biodiversity. Compared with the issue of climate change, where even a radical shift in local policy makes only a small imprint on the overall output of emissions, these problems *are* possible to tackle in a way that makes local influence visible.

The European Commission has shown growing interest in creating sustainable mobility. The Fifth Environmental Action Programme, Green Papers, the Amsterdam Treaty, and the Sixth Environmental Action Programme all emphasize that environmental concerns should be integrated in transport policy. Most of the EU Member States have engaged in integrating environmental aspects in the transport sector focusing upon the issue of interregional transport for work place commuting and delivery of food and energy. However, narrowing the perspective to transport that takes place within the city – which has been the topic of this chapter – there seems to be ample room for a local approach to local problems caused by urban transports.

Time for innovations

The average speed of vehicles in major European cities has declined by 10 per cent over the last 20 years. Thus, there is a need to create a more efficient urban transport system, and there is also a strong economic rationale for doing that. The external cost of road traffic congestion within the EU alone amounts for 0.5 per cent of GDP, and traffic forecasts for the next ten years show that if nothing is done the cost will increase by 142 per cent to reach 80 billion euro a year. Traffic accidents cost 2 per cent of GDP, households annually spend 600 billion euro (14 per cent of each household's annual income) on transport, and 70 billion Euro are invested annually in transport infrastructure (1 per cent of GDP). Traffic congestion, together with delays in flights and bottlenecks in the railway system, result in a consumption of an extra 1.9 billion litres of fuel, which is some 6 per cent of annual consumption (IP/01/1263).

If no change in the urban transport system take places, the future development will lead further away from and not towards global as well as local sustainability. As declared by Loyola de Palacio, Vice-President responsible for energy and transport within the EU: 'Only new approaches will enable us to deal successfully with the growth of pollution and congestion caused by transport in cities' (EU, 2001). New roads and less fuel-consuming cars cannot be the solution. Instead there is a need for realising new ideas in practice, i.e. solutions that do not only include efficiency of the existing transport system, but also change of transport modes.

The need to connect local innovations with other policy levels

On the macro-level, there are few signs that the development within the EU goes in the right direction. The structural funds – representing more than 35 per cent of the total budget for the EU and playing a major role for the development of Europe's transport infrastructure – have so far more or less neglected environmental considerations in their activities. Nevertheless, they have a great potential for contributing to sustainable mobility. On the local level a number of initiatives have been conducted showing that there are opportunities for cities constructively to deal with the urban transport challenge. For example, the Car Free Cities slogan – 'In town without my car' – is a symbolic gesture indicating a will to restructure the urban transport system in an eco-friendly direction. In contrast to many other environmental problems, urban transport offers opportunities for cities to develop pilot projects on urban mobility without co-ordinating with actors at higher levels.

If this kind of micro-innovation is to influence macro-development, however, it is important that mechanisms are created which connect these projects with the wider context. Otherwise these examples may serve the function of greenwash – good practices but serving to legitimate the existing unsustainable transport system. A continued increase in transport may result in a situation where more eco-friendly transport modes increase relative to other modes and more cities conduct projects for sustainable mobility – at the same time as urban transport by car increases in absolute terms. In that case, best practices run the risk of being part of a Pyrrhic victory; battles fought successfully on the local level, but yet yeilding defeat on the European level.

Micro-innovation can even be framed in a way that does not only challenge other cities, but also national and regional levels, thereby stimulating a broader change of urban transport policies in society. From this perspective, it is important that cities do not underperform, but use their space for action to develop and test innovative urban transport alternatives. The dissemination of results – positive and negative – to the wider community as well as formulating demands to higher political levels are important policy tools. However, emphasising the importance of local innovations runs the risk of neglecting the responsibility of the supra-local levels – national and regional. If the development on the local level is not backed up by relevant decisions at national and regional levels, there is a risk that government initiatives become restricted to the financing of research and pilot projects. From a local perspective it is not enough that the EU allocates some *ad hoc* resources and creates mechanisms for the transfer of practical knowledge. Aside from its role as a facilitator of learning processes, the critical question for the future will be to what extent will the EU integrate environmental concerns in its urban transport policy?

To summarize, urban transportion produces an enormous loss of valuable time through traffic congestion, unhealthy – psychological as well as medical – environments for urban inhabitants, and contributes substantially to climate change. The EU is well aware of this problem, at the same time as it views mobility as an unquestioned goal. By the concept 'sustainable mobility' it hopes to reconcile and achieve a number of transport related goals, i.e. a safe, efficient, competitive, socially and environmentally friendly common transport policy. However sensible these goals may be from a sustainability point of view they unfortunately run the risk of being wasted by aggressive demands for accelerated economic growth of the EU economy. Ironically, these demands are strongly raised also from within the EU itself.

Notwithstanding this complication, there is also room for cautious optimism. As argued by Tim Hart in a recently published review article on transport and the city: '...with a favourable wind from public opinion...urban vehicle miles by car will stop rising within the next ten years and then begin to fall' (Hart, 2001: 120).

Notes

1. One euro = US$0.866 (February 2002).
2. The indicator-based reporting mechanism is known as the Transport and Environmental Reporting Mechanism (TERM). TERM is a joint initiative between the Commission, the Statistical Office of the European Communities (Eurostat) and the EEA, and is viewed as an important tool in furthering the integration process by moving beyond primarily descriptive indices and addressing the performance and efficiency of transport systems. TERM incorporates measures such as transport pricing, land use planning and economic performance in order to assess the state of policy action.
3. Term 2001, a report by the European Environment Agency (EEA), has recently sounded the alarm bell. Pollution caused by the transport sector threatens to compromise Europe's commitments entered into at Kyoto, namely an 8 per cent reduction in greenhouse gas emissions between 2008 and 2012 (RTD info, 2001, No. 32, p. 28).

14
Managing Transport Demand in European Countries

John Whitelegg with Nicholas Low

Introduction

It is coming to be accepted that the growth in demand for transport must be 'decoupled' from economic growth (EC, 2001). The future direction of transport policy should be able first of all to reduce the rate at which car-kilometres and tonne-kilometres of freight transport grow, and then reduce these in an absolute sense. This can be achieved without sacrificing the prospect of economic growth. Indeed many transport commentators in Europe would argue that we can only continue to develop our economies if we solve traffic congestion and pollution problems. This European tendency echoes changes in UK national transport policy which, in spite of some residual support for large and expensive new roads, is nevertheless oriented in the direction of demand management and reducing the need to travel. This represents a fundamental policy shift that is more demanding culturally and psychologically than it is in any technical sense. The policy shift, however, is real and documented in UK governmental statements especially the influential 'PPG13' (Planning Policy Guidance Note 13, Transport). But the views of many politicians and engineers is that only roads can bring relief from congestion and only roads can deliver economic gains for hard pressed regions. In this chapter we review the power of demand management solutions to solve both sets of problems.

Transport policies in European countries

Lidskog, Elander and Brundin (Chapter 14) reviewed transport policy at European Union level. Here we review transport policies in specific European national contexts, with particular emphasis on demand management. A number of policies are either currrently in use or actively under development in European countries to solve transport problems. They include:

- urban road pricing (including congestion pricing);
- fuel taxation;

- taxation of parking spaces at the workplace;
- taxation of parking spaces at car-intensive developments (e.g. out-of-town and edge of town retailing complexes, airports, leisure centres, sports facilities);
- regional norms on car parking provision to deter a competitive bidding-up process;
- financial support from hypothecated revenues for quality public transport, coordination, integration, dense pedestrian and cycle networks and innovative programmes of accessibility enhancement for rural areas;
- substantial policy integration at the national level so that transport and land use planning policies support health policies, climate change policies and vice versa;
- modification of the 'predict and provide' approach which still determines policies towards airport capacity and housing provision;
- the establishment of regional transport authorities following the German models or the Danish model (e.g. the Rhine-Main transport authority in Frankfurt, HUR in Copenhagen – these authorities bring public and private finance to bear on the supply of public transport, high quality integration and co-ordination, 'environment ticketing' schemes and high quality information);
- providing new methods of funding public transport e.g. fuel taxation as in Germany and employer contribution as in Paris (the 'versement' tax on employers);
- eliminating subsidies to private motorized transport through the company car, business mileage and corporation tax regimes.

Not all these policy areas will be discussed here. We focus on those relating mostly to demand management and the provision of alternative transport options.

Road pricing and fuel taxation

These sources of revenue provide an attractive policy option, particularly if the funds are recycled into the local economy to support all the alternatives to the private car. According to the OECD (1995) survey of transport policy options road pricing is being considered in some shape or form in most OECD countries. The Mayor of London intends to introduce a cordon based system in 2003 which will require vehicles to pay UK£5 to enter the charging area (very roughly coinciding with central London). Similar plans are well advanced in Cambridge and Edinburgh (UK), toll systems on roads exist in Norway, and Stockholm is planning to introduce such a system.

Road pricing is generally suggested for those locations where the growth rate in traffic is already *the lowest* across geographical situations. Thus the growth of traffic into and out of central London has been far lower than the growth in outer London or the growth on the M25 corridor. Road pricing is best seen as a strongly supportive measure alongside a battery of other measures including strong land use controls and modal preference restructuring.

The view of the OECD (1995: 154) is that 'The key to the sustainable development strand is a substantial and steadily increasing fuel tax coupled with (other) measures'. The UK had a policy of increasing fuel tax by 6 per cent above the rate of inflation at each annual budget, but this has now been abandoned by the Labour government. The OECD suggests that the impact of a 7 per cent per annum rise in fuel costs in real terms would be to quadruple fuel prices in 20 years. This would eventually lead to lower car ownership levels compared with what they would otherwise be, fewer car trips and shorter trip lengths. An overall reduction in car-trip making of about 15 per cent, a reduction in trip length of about 25 per cent, and an overall reduction of vehicle kilometres of one third is predicted if fuel prices rise by a factor of 2.5 (OECD, 1995: 156).

Denmark has a very high tax on the purchase of new cars (160 per cent) and has one of the lowest levels of car ownership in the European Union. The Stockholm proposals provide a model for UK local authorities. Stockholm will be divided into ten zones covering the whole of the built up area, served by 90 fee stations. Light vehicles (e.g. cars) would pay 0.45 or 0.55 Euro per transit on weekdays between six o-clock in the morning and seven in the evening. The lower charge is for automatic debiting and the higher for manual systems. Heavy goods vehicles would pay 1.1 Euro per transit if fitted with noise reduction technology and 1.4 Euro if not. Once again higher charges would apply to manual systems. The differential charge for noise indicates a real environmental benefit from road pricing. Vehicles can be charged on a number of different noise and pollution criteria to help achieve air and noise quality objectives as well as congestion targets. The Stockholm scheme is estimated to bring in about 140 million Euro a year. On the expenditure side, 13 per cent is allocated to administrative costs, 79 per cent is refunded to residents and the rest set aside for noise reduction and public transport expenditures (EFTE, 1996).

Parking taxes

Free car parking (as in local government or employer provided spaces) is a strong incentive to use the private car for commuter purposes. A well developed system of public and private car parking charges already exists in most Euroepan countries. In the UK, however, more than half of the available car parking in a UK town or city (though it varies with the location) is 'free parking'. Similarly, private non-residential car parking (PNR) at workplaces, hospitals, universities and airports provides a powerful incentive for the use of car-based transport for commuting and other purposes and for the use of cars in the course of work.

If local authorities are going to be successful in achieving traffic reduction targets there will need to be strong disincentives to add to the supply of PNR spaces and strong incentives for employers and site managers to develop mobility options that give far more choice than the car. This will in turn have an impact on the initial location choice of businesses and residences to the benefit of public transport, walking and cycling options. A specific car space tax is suggested as a clear fiscal measure to achieve these objectives.

Parking restrictions

Restrictions in mainland European cities such as Zürich and decisions, as in Amsterdam, to reduce car park numbers provide best practice examples (Lemmers, 1994). Good practice parking policies exist in the UK in Sheffield, Winchester, Leeds, Southampton, Cambridge and Edinburgh. A MVA consultancy study of Bristol for the Department of Environment, Transport and the Regions (DETR) shows a 75 per cent reduction in on-street parking. Higher charges and enforcement of planning permission for PNR parking will reduce car trips into central Bristol by 41 per cent (T 2000, 1997a).

Reallocating spatial and modal preferences

There are many isolated examples of successful policies in this area: the Manchester Metrolink, bus lanes in several British cities, and Maidstone's (UK) Integrated Sustainable Transport (MIST) project, Zürich's prioritization of public transport, car-free residential and city centre areas (Lübeck, Amsterdam, Berlin, Edinburgh), building homes on former car parks; bicycle priority schemes and planning in York and Cambridge (UK), Delft and Groningen (the Netherlands), Detmold and Rosenheim (Germany), Copenhagen's cycling strategy, and Darmstadt's (Germany) encouragement of cyclists and pedestrians to share the same large car-free space in the city centre, SMART buses in Liverpool, new tram systems in Strasbourg, innovative car-sharing initiatives (StattAuto) in Berlin, Bremen and Edinburgh. 3000 participants in the Berlin car sharing scheme have removed 2000 cars from the roads of Berlin. Vienna has adopted a policy of constructing several hundred extended pavements at crossings and tram stops to improve the safety of pedestrians.

Traffic management

Groningen (Netherlands) has developed a sector access model; Bochum (Germany), has prioritized its trams in preference to cars, Göteborg (Gothenburg Sweden), has divided the CBD into five cells which has had the effect of reducing car mobility by 50 per cent. Houten (Netherlands) with a population of 30,000 has given preference to bicycles, restricting access by sectors and imposing traffic restraint. Oxford (by means of parking controls and a Park-and-Ride scheme) over the last 20 years has delivered one of the lowest rates of city centre traffic growth of any UK city.

Marketing strategies

Large scale marketing exercises have increased bus patronage in Lemgo (Germany). Similar but less ambitious schemes can be found in the UK, e.g. SMART buses (Liverpool) and TravelWise schemes. System-wide, discounted tickets have helped increase public patronage in Germany, e.g. the *Umweltkarte* (environment tickets). In Freiburg a reduction of 4,000 cars per day on the roads to the city centre is attributed to the introduction of the *Umweltkarte*. The German national rail system has increased its patronage through the introduction of the *Bahnkarte* system which provides 50 per cent discounts on all rail ticket purchases after the acquisition of an annual card costing DM250.

Demand management projects

There is already much experience of significant modal shifts and associated traffic reduction in European cities and regions. For instance Lemgo increased bus usage from 40,000 to over one million in one year. Zürich has held levels of car ownership and traffic volumes constant for a decade whilst public transit use has soared. And Houten has developed a comprehensive bicycle–pedestrian network and has cut car trips per household by 25 per cent (Hammond, 1994).

Swiss and German research on car-sharing shows that people who have joined a car-sharing scheme (not car-pooling) and who have previously owned a car have reduced their car mileage by 50 per cent. The Federal Ministry of Transport in Germany estimates that car sharing will reduce annual vehicle kilometres by 7 billion in Germany. While In Europe the figure is put at a reduction of 30 billion vehicle kilometres (EC, 1997).

In Zürich substantial investment in public transport coupled with parking and access policies have led to the stabilization (but not reduction) of traffic growth and to an increase of 30 per cent in the use of public transport (1985–90). The level of use of public transport is now 470 trips per inhabitant per year, about twice the level of comparable European cities (OECD, 1995). In Aachen (Germany) traffic flowing into the city centre has been reduced by 85 per cent over ten years, the car's share of transport has gone down from 44 per cent to 36 per cent and NO_x pollution has been reduced by 50 per cent (Poth, 1994). In Bologna, Italy, between 1982 and 1989 a deliberate policy of traffic restraint, closing streets and providing park-and-ride facilities produced a 48 per cent drop in motorized traffic entering the historic core and a 64 per cent drop in numbers of cars. In Groningen (Netherlands), thanks to a co-ordinated policy of traffic restraint and provision for alternative modes, in 1990 48 per cent of all trips within the city were by bicycle, 17 per cent on foot, 5 per cent by public transport and just 30 per cent by car (Pharoah and Apel, 1995).

Traffic reduction and alternative transport strategies have been adopted successfully in UK towns. For instance in Manchester the Metrolink tram has taken up to 50 per cent of car journeys off roads in the area it serves. It has replaced over one million car journeys into the city centre each year.

Five per cent of car users switched to a new 'City Express' bus service in Belfast in the first 6 months of operation, while Edinburgh has set itself a traffic reduction target of 30 per cent. In Leicester 10 per cent more 7–9 year olds were allowed to walk to school after traffic calming (T 2000, 1997b). Levels of cycling have more than doubled in one of the 'Safe Routes to Schools' pilot projects even without the necessary infrastructure works being carried out. More than 120 pupils at Horndean Community School in Hampshire are regularly cycling to school compared with about 50 last autumn and just 36 when the project began at the end of 1995 (Network News, 1997).

Integrated travel across all public transport modes is also important. The 'Carte Orange' in Paris covering all modes of transport, introduced in 1975, led to a 36 per cent increase in bus patronage. The London travel card led to a 16 per cent increase in public transport use at a time of decline elsewhere (EC, 1996a).

Green Commuter strategies on the part of particular organizations are increasingly common in the UK, for example Nottingham (City of Nottingham, County of Nottinghamshire, Queens Medical Centre, the Universities and Boots Ltd.), Derriford Hospital (Plymouth), Oxford University (a planning agreement). Pfizer, the US based multinational company has reduced car use at its Sandwich (Kent, UK) plant by 12 per cent in the three years of operation of a transport plan. The Rijnstate Hospital in the Netherlands has restricted its car parking provision to 400 spaces for 2,050 staff. Transport Demand Management policies have increased the use of public transport from 8 per cent to 40 per cent of all journeys. Restricting car parking availability was the key to this success.

Traffic reduction for heavy goods vehicles

Heavy goods vehicles (HGVs) are a long standing problem in towns and cities, on trunk roads through villages and in or near national parks. They pose a very serious problem on Alpine transit routes and on key sections of motorways in Germany, France and the UK. In general their impact is much greater than their numbers would suggest. Their impact on noise, road damage, air quality and the fears of pedestrian and cyclists is large and there is a strong case for reduction in ways that can protect the economy of towns and cities and the consumer who has come to depend on goods and services supplied by HGVs. Considerable progress has been made in this area in mainland Europe, particularly Germany, whilst hardly any progress at all has been made in the UK. In Germany HGV reduction strategies which pay attention to the commercial interests of the companies involved are generally referred to as 'City-Logistik' strategies. Eliminating unnecessary freight movement can also mean reductions in costs.

City Logistics involves setting up new partnerships and styles of co-operation between all those involved in the logistics chain and in delivering or receiving goods in city centres. These partnerships offer significant reductions in vehicle kilometres and truck numbers and are currently in existence in Germany and Switzerland. City Logistics are a very clear illustration of the importance of developing high quality organizational arrangements and inter-company co-operation agreements in addition to whatever new technology might be appropriate. City logistics have taken transport operations into an area of development that builds links and emphasizes co-operation across all players and interest groups.

In Germany these *City Logistik* partnerships are in operation in Berlin, Bremen, Ulm, Kassel and Freiburg. The Freiburg example offers several pointers to the future shape of freight transport in urban areas. There are currently twelve partners in the scheme. Three of the partners leave city centre deliveries at the premises of a fourth. The latter then delivers all the goods involved in the city centre area. A second group of five partners delivers all its goods to one depot located near the city centre. An independent contractor delivers them to city centre customers. A third group, this time with only two service providers, specializes in refrigerated fresh products. These partners form an unbroken relay chain, one partner collecting the goods from the other for delivery to the city centre.

The Freiburg scheme has reduced total journey times from 566 hours to 168 hours (per month), the monthly number of truck operations from 440 to 295 (a 33 per cent reduction) and the time spent by lorries in the city from 612 hours to 317 hours (per month). The number of customers supplied or shipments made has remained the same. The Kassel scheme showed a 70 per cent reduction of vehicle kilometres travelled, and 11 per cent reduction in the number of delivering trucks. This has reduced the costs of all the companies involved and increased the amount of work that can be done by each vehicle/driver combination.

There are clear economic benefits arising from lorry traffic reductions. These reductions have benefited the companies through higher levels of utilization of the vehicle stock. It is not in the interests of logistic companies to have expensive vehicles clogged up in city centres, one way systems and on circuitous ring roads.

Traffic reduction in rural areas

Rural areas have higher levels of dependency on cars than urban areas and have experienced a steady decline in recent years in the range and quantity of facilities that represent the normal everyday destinations for our trips. The decline in rural shops, post offices, schools and health care facilities has been documented in most of the UK's rural areas. For these reasons special care is needed with traffic reduction policies in rural areas.

Rural areas, however, are not universally perceived as particularly difficult in terms of public transport provision and facility development. Rural transport and facility density in Switzerland and Norway are well developed and sit amongst a number of other measures designed to support rural residents. In the UK this support network is lacking and it would be unreasonable to expect transport policies to make up for the huge deficits in other policies. The existence of a 'rural transport problem' is largely dependent on the extent to which organizational and fiscal changes over the last 30 years have left rural areas unsupported. When this support is restored, e.g. through financial incentives that will support small schools, post offices, shops and rural enterprise as well as affordable housing then the 'rural transport problem' is rendered less intractable.

Central government can support rural areas through a policy of providing resources for small facilities in a dispersed pattern in a rural area, particularly in education and health care. In the UK local authorities can support rural areas through the provision of the 1988 Local Government Finance Act. This already happens in East and West Sussex where all but one of the districts offer rate relief to village shops.

Rural inhabitants will still need to travel and unless transport initiatives are vigorously pursued this is likely to be by car. However, there are potentially a number of alternatives to the car in rural areas:

- much improved bus services on main routes into larger settlements;
- improvements to rural rail services where these exist;
- community/voluntary car schemes;
- community bus/dial-a-ride schemes;
- improvements to pedestrian and cycling facilities;
- shared use of vehicles, e.g. post buses;
- home deliveries.

The exact mix of transport opportunities will vary from area to area and from the deep rural situation to circumstances where a large market town is accessible within half an hour by bus. UK experience with rural bus services and community bus services up to 1985 was successful in many places but was dealt a severe blow by bus deregulation in the 1985 Act. This Act is in urgent need of reform to encourage innovative, community-run bus services in rural areas.

In spite of this unhelpful public transport regime there are still very good examples of high quality bus services in rural areas. The bus services in Cerrig-y-Druidion in North Wales provide such a link (to Ruthin and Denbigh) and are well used. Rural railways also continue to provide good quality public transport in those areas that are still served by this mode. Recent research by TR&IN in Huddersfield has shown how rural lines currently serve their populations (Exeter to Barnstaple, Derby to Matlock, Ipswich to Suffolk and Huddersfield to Sheffield) and how they could do much more to offer a high quality, affordable alternative to the car (Chapter 9 of this volume by Takeda and Mizuoka shows the deleterious effect on rural transport in Japan of privatization of the rail system).

In rural Oxfordshire a study of Cholsey and Chalgrove villages showed that residents of the village with the poorer level of public transport (Chalgrove) travelled 30 per cent further by car than Cholsey residents. Cholsey has a bus or a train at least every hour. The survey (Environment Change Unit, Oxford University, 1996) also showed that the average distance for car journeys within both villages was one mile or less, indicating a significant potential for transfer to non-car modes. In Germany the 'Bürgerbus' initiative has set a high standard for affordable, frequent, community managed rural bus services. These buses have been funded by the State government of North Rhine Westphalia and are operated by locally managed companies. They cover a network of market towns and sparsely populated areas at a variety of frequencies and carry between 2,000 and 18,000 passengers per annum.

It would be a mistake to assume that rural transport demand is dominated by large numbers of long journeys in situations where there is no public transport. The reality is far more varied and has considerable potential for intervention to bring about a shift away from the car.

The economic consequences of traffic reduction

There is a well-developed literature on this subject. Most of the work is German in origin where resources have been devoted to empirical research on the relationship between traffic restraint (e.g. reducing the number of car parking places in cities) and retail viability. The findings of this research, carried out by the German Institute for Urban Research in the late 1980s and early 1990s, are very clear: 'A study in Germany suggests that retail trade in central city districts increases with policies that encourage environmentally friendly transport modes. Of the 38 cities studied, 14 had above average retail growth. Of these 14, ten had below average provision of infrastructure for the car' (EC, 1996b: 178). This is not really surprising. There is a large literature on the costs of congestion and the scale of the defensive expenditures that have to be deployed to cope with the air pollution, noise pollution, road traffic accidents and congestion impacts of traffic growth and traffic concentration in space and time (Maddison *et al.*, 1996).

Authoritative European surveys agree that the total external costs of transport in 17 European countries amount to 270 billion Euro per year, an average of 4.6 per cent of GDP. The road total is 50 times higher than the rail total and for all practical purposes walking and cycling can be regarded as having zero external costs. The full implementation of already accepted EU policy for internalizing the external costs of transport would significantly reduce the number of vehicle kilometres of car and lorry travel while at the same time expanding the use of other modes and liberating significant resources for investment in social infrastructure (e.g. education and training), environmentally high performing buildings, and innovation in design and manufacturing to enhance the international competitiveness of UK businesses.

In a seminal study of Japan's urban transport system and economic performance, Hook (1994) associates Japan's reliance on non-motorized transport and rail transit with its economic success:

> High urban density and a transportation system heavily reliant on non-motorised transport and its linkages with rail based mass transit have been critical to Japan's economic success. By minimizing aggregate transportation costs, Japan has been able to minimize its production costs, making its goods more competitive in international markets. Further, by discouraging the consumption of private automobiles and encouraging savings, a larger pool of potential investment capital was created, also critical to rapid economic growth... the automobile far from being a symbol of economic prowess is more a symbol of economic assets being wasted on consumption instead of on job creating and productivity-increasing investment. Meanwhile the bicycle and other non-motorised vehicles, far from being a symbol of economic backwardness, are more symbols of a society able to meet its passenger transport needs in the most cost effective and least environmentally damaging way, allowing scarce economic resources to be invested elsewhere. (Hook, 1994)

It is perhaps significant that Japan's economic woes have in no way been alleviated by a movement towards private car transport and privatization of the railways, and they may have been exacerbated (see Chapter 9 of this volume).

There are considerable benefits to be had from public transport investments. Steer *et al.* (1997) in their report for Transport 2000 show that the total non-user benefit from investing in the Midland Metro Line 1 amounts to UK£112.85 million at 1989 prices. Evidence from Portland, Oregon (USA) shows a major wave of economic revitalization from the new transit system and its associated land use planning (Centre for Clean Air Policy, Washington DC, 1997). Portland's economic decline in the 1960s and 1970s was reversed by new high density housing in the downtown area, conversion of streets to pedestrian-friendly configurations, replacing a riverside motorway with an esplanade, stringent parking restrictions, free public transport in the central area using a new light rail system and the scrapping of road schemes. The result has been a revitalized city centre with 30,000 more jobs and 40 per cent of commuters using public transport.

Detailed empirical research in Germany shows that there is no relationship between the amount of car parking provision in the main city centres and the amount of retail spending in those areas (Baier and Schaefer, 1997). Freiburg, with very low numbers of car parking spaces per inhabitant, has a higher level of retail spending than Wetzlar with four times the number of spaces per inhabitant than Freiburg. In the case of public transport there is a very strong relationship. The higher the number of public transport arrivals in the cities the *greater* is the level of retail activity in those centres. The evidence in Europe points to the economic success of traffic restraint policies and to the non-existence of damaging economic consequences.

The literature and experience from all advanced industrial countries that have invested in alternatives to the car and in forms of mobility other than the car shows that there are measurable economic benefits and gains from doing so. Traffic reduction is not about stopping people travelling. Transferring trips to modes of transport other than the car or planning for land use arrangements and accessibility patterns that stimulate innovations in supply methods are more likely to create jobs than to destroy them. Indeed sustainable transport policies with traffic reduction at their heart are examples of strategies that have the potential to create real lasting local jobs that can sustain local communities at a time when globalization tendencies are making jobs far more mobile than at any other time in the past 50 years.

Concluding remarks

Achieving significant modal shifts, shifting people out of cars, reducing freight transport by eliminating unnecessary movement, creating more live-able cities, towns and rural areas and meeting traffic reduction targets is entirely feasible. The relative lack of progress in the UK, Spain, Portugal and Greece in recent years is not indicative of psychological obstacles or even a carefully balanced choice to go for cars and total freedom of movement. It is the result of policy that either intentionally or unintentionally has encouraged the growth in demand for motorized transport.

This stance also implies a corollary. Moving away from car-dependence and shifting to lower levels of car use with higher levels of use of alternative transport brings with it multiple benefits: it brings economic gains to city centre retailing, it brings *national economic benefits* – by re-invigorating neighbourhoods and communities, improving health and reducing the costs of dealing with the sickness effects of transport, and reducing the total amount of public expenditure devoted to new infrastructure. It helps to achieve *reductions in greenhouse gases*, enabling countries to meet their global obligations. It is *inclusive and assists social sustainability*, helping the young, the disabled, the elderly and the poor.

Traffic reduction involves new ways of designing and implementing policy as well as new policy objectives. These new ways of working will require clear action by national government to put the right conditions in place that will allow local authorities to do their work effectively. It will also mean highly co-ordinated and integrated strategies at the local authority level to bring together the traditional transport, highway and planning functions with the agents of land use change and traffic generation. A collaborative model of working will have to be formulated and implemented with the specific aim of reducing the need to travel and shifting the pattern of transport away from road use. This will mean the development of integrated transport–land use policy in which the first consideration is not building new infrastructure but planning liveable and environmentally sustainable cities.

15
Lessons from Asia on Sustainable Urban Transport

Paul Barter, Jeff Kenworthy and Felix Laube

Introduction

Urban transport, motorisation and the development of 'automobile dependence' have become critical factors in the future liveability of cities, not least those in Asia where motorization is reaching an ever wider range of cities. Urban residents and policy makers struggle with the escalating impacts of private transport and how best to provide for people's transport needs in cost-effective and more sustainable ways. These efforts are part of the wider quest for sustainable, liveable and equitable cities across a broad range of factors, many of which are affected to some degree by the nature of the transport system.

This chapter confronts these large issues with a policy-oriented discussion that focuses on Asian cities in an international perspective. The comparisons are informed by a large data set (discussed below) as well as by earlier investigations by the authors into a subset of these cities (Barter, 1999; Kenworthy and Laube, 1999). The chapter focuses on the regions of East Asia, Southeast Asia and South Asia.[1] This group of regions provides an interesting 'laboratory' on urban transport where we find a host of variations in urban transport patterns.

There are numerous themes that we could explore but in the limited space available we will focus on certain lessons from the past experiences of Asian cities that now have high or middle incomes and which are especially relevant for low-income cities. Two themes arise strongly from this focus, both relating to the relative priority given to different modes of transport. They are: priorities in investment between the main modes of passenger transport (public transport, private transport and non-motorized transport); and policies affecting the pace of motorization and the growth of private vehicle use.

Data on urban transport from a large sample of cities

The set of data referred to in this paper is drawn primarily from the *Millennium Cities Database for Sustainable Transport* (Kenworthy and

Laube, 2001), which was compiled by the authors over three years for the International Union (Association) of Public Transport (UITP) in Brussels. The database provides data on 100 cities in all continents. Data summarized here represent averages for nine groups of cities from 84 of the fully completed cities (listed in Table 15.1).

A detailed discussion of methodology is not possible in this chapter. The database contains data on 69 primary variables, which can mean up to 175 primary data entries. The methodology of data collection for all the factors

Table 15.1 Urban areas in the millennium cities database for sustainable transport[b]

Western Europe (WEU)			**United States of America (USA)**	**Canada (CAN)**
Munich	Copenhagen	Berlin	S. Francisco	Vancouver
Frankfurt	Stockholm	London	Washington	Calgary
Zurich	Ruhr	Barcelona	New York	Toronto
Geneva	Nantes	Madrid	Denver	Ottawa
Düsseldorf	Graz	Glasgow	Chicago	Montreal
Bern	Marseilles	Manchester	Atlanta	
Lyon	Helsinki	Newcastle	Houston	**AUST/NZ (ANZ)**
Paris	Amsterdam	Athens	Los Angeles	Sydney
Stuttgart	Brussels		Phoenix	Perth
Vienna	Bologna		San Diego	Melbourne
Oslo	Rome			Wellington
Hamburg	Milan			Brisbane
High income[a] Asia (HIA)	**Middle income[a] Asia (MIA)**	**Low income[a] Asia (LIA)**	**Middle income[a] Other (MIO)**	**Low income[a] Other (MIO)**
Tokyo	Taipei	Guangzhou	Tel Aviv	Bogotá
Osaka	Seoul	Shanghai	Prague	Teheran
Sapporo	Kuala Lumpur	Manila	Curitiba	Tunis
Hong Kong	Bangkok	Jakarta	Riyadh	Cairo
Singapore		Beijing	Budapest	Dakar
		Ho Chi Minh City	Sao Paulo	Harare
		Mumbai	Johannesburg	
		Chennai	Cape Town	
			Krakow	

Notes

[a] For the purpose of grouping these cities, the cut-off points in terms of Gross Regional Product (GRP) per capita (1995 prices) between high-income and middle-income cities and between middle-income and low-income cities have been chosen to be US$16,000 and US$3,000 respectively.

[b] The following cities are also included in the database but unfortunately could not be included in the analysis here due to incomplete data sets: Lille, Turin, Lisbon, New Delhi, Buenos Aires, Rio de Janeiro, Brasilia, Salvador, Santiago, Mexico City, Caracas, Abidjan, Casablanca, Warsaw, Moscow, Istanbul.

was strictly controlled by agreed-upon definitions contained in a booklet of over 100 pages. Data were carefully checked and verified before being accepted into the database. From this complex range of primary factors, some 230 standardized variables have been calculated addressing a wide range of transport-related issues. For this chapter only a selection of salient features are chosen for comment. The data are for the year 1995 (although in certain cities the reference year is a year close to 1995). Data collection commenced in 1998 and was completed at the end of 2000. Currently, data for 1995 provides the latest perspective one can reasonably expect for an urban study of this magnitude.

Table 15.2 presents relevant data for the nine groups of urban areas. These particular groupings of cities were chosen with the help of several applications of hierarchical cluster analysis. This revealed that the regional groups, USA cities, Canadian (CAN) cities, Australian and New Zealand (ANZ) cities and Western European (WEU) cities generally corresponded with clusters of cities in the data set, suggesting that using these regional groupings would not be misleading. The cluster analyses also found that most of the Asian cities consistently fell into a number of clusters each of which were substantially Asian in membership. However, these Asian clusters generally did not follow sub-regional boundaries. Clusters among the remaining cities (East, Southeast and South Asia) were not obviously regional (or even sub-regional) in nature either. Further exploratory investigation of the data set suggested that a combination of income-based and regional groupings was the best option for comparing groups of 'non-Western' cities in the sample.

The Asian cities, therefore, were placed into three groups – High Income Asian (HIA), Middle Income Asian (MIA) and Low Income Asian (LIA) cities. The remaining cities in the sample were placed into two groups, Middle Income Other (MIO) and Low Income Other cities (LIO) as shown in Table 15.1. The choice of cut-off points between the higher-income, middle-income and lower-income groups was influenced by the cluster analyses and other exploratory analysis. For example, the choice of a high cut-off between middle-income and high-income cities allows Taipei to be grouped with Bangkok and Kuala Lumpur, with which it was consistently grouped by the cluster analyses.

Asian cities in international context

In this section, we first examine some background issues before moving on to the policy-related issues that are the main focus of the chapter. All references to group averages refer to Table 15.2. Data on individual cities are from Kenworthy and Laube (2001).

Land use characteristics

A striking feature of the Asian cities as a group is their high density.[2] The average urban densities of the three Asian groups (ranging from 134 to 206

Table 15.2 Land use and transport system characteristics by groupings of cities, 1995

		USA	ANZ	CAN	WEU	HIA	MIA	LIA	MIO	LIO
Land use and wealth										
Urban density	persons/ha	14.9	15.0	26.2	54.9	134.4	164.3	205.6	53.7	122.1
Proportion of jobs in CBD	%	9.2%	15.1%	15.7%	18.7%	20.1%	13.1%	31.8%	16.8%	21.2%
Metropolitan gross domestic product per capita	USD	$31,386	$19,775	$20,825	$32,077	$34,797	$9776	$1689	$6625	$1949
Transport investment cost										
% of metro. GRP spent on public transport investment	%	0.18%	0.30%	0.18%	0.41%	0.47%	1.22%	0.53%	0.39%	0.62%
% of metro. GRP spent on road investment	%	0.86%	0.72%	0.87%	0.70%	0.96%	1.34%	1.82%	0.70%	0.75%
Private transport infrastructure indicators										
Length of expressway per person	m/person	0.156	0.129	0.122	0.082	0.022	0.027	0.004	0.043	0.009
Parking spaces per 1000 CBD jobs		555	505	390	261	121	164	55	374	134
Public transport supply and service										
Public transport seat kilometres of service per capita	seat km/ person	1556.8	3627.9	2289.7	4212.7	5535.2	2734.4	2057.4	3282.8	3322.2
Rail seat kilometres per capita (Tram, LRT, Metro, Sub. rail)	seat km/ person	747.5	2470.4	676.4	2608.6	2719.9	361.8	250.0	1683.6	120.4
% of public transport seat km on rail	%	34.2%	65.2%	27.8%	55.5%	57.2%	13.1%	12.9%	33.6%	10.1%
Overall average speed of public transport	km/h	27.4	32.7	25.1	25.7	33.2	16.4	16.6	24.8	21.1
Ratio of public versus private transport speeds		0.58	0.75	0.57	0.79	1.08	0.78	0.80	0.70	0.71
Private transport supply (cars and motorcycles)										
Passenger cars per 1000 persons		587.1	575.4	529.6	413.7	217.3	198.3	38.0	265.1	71.2
Motor cycles per 1000 persons		13.1	13.4	9.5	32.0	65.8	154.0	95.6	14.7	15.1

Table 15.2 Continued

		USA	ANZ	CAN	WEU	HIA	MIA	LIA	MIO	LIO
Mode split of all trips										
* non motorized modes	%	8.1%	15.8%	10.4%	31.3%	29.1%	19.8%	50.1%	27.9%	36.3%
* motorized public modes	%	3.4%	5.1%	9.1%	19.0%	32.3%	25.6%	28.3%	26.6%	32.8%
* motorized private modes	%	88.5%	79.1%	80.5%	49.7%	38.6%	54.6%	21.6%	45.5%	30.9%
Private mobility indicators										
Passenger car passenger kilometres per capita	p.km/ person	18,155	11,387	8645	6202	3724	3517	785	4133	1172
Motor cycle passenger kilometres per capita	p.km/person	45	81	21	119	100	1165	416	78	90
Public transport mobility indicators										
Total public transport boardings per capita	bd./person	59.2	83.8	140.2	297.1	464.1	274.2	267.3	340.5	234.4
Rail boardings per capita (Tram, LRT, Metro, Sub. rail)	bd./person	21.7	42.5	44.5	162.2	284.8	38.9	30.0	159.0	15.6
Proportion of public transport boardings on rail	%	25.7%	48.8%	28.9%	50.0%	62.0%	12.8%	11.0%	33.1%	7.6%
Proportion of total motorized passenger km on public transport	%	2.9%	7.5%	9.8%	19.0%	50.3%	26.9%	51.1%	36.6%	54.2%
Public transport productivity										
Public transport operating cost recovery	%	35.5%	52.7%	54.4%	59.2%	138.5%	98.8%	138.6%	82.9%	107.9%
Overall transport cost										
Total passenger transport cost as % of metro. GRP	%	11.79%	13.47%	13.72%	8.30%	5.41%	13.60%	13.63%	15.45%	17.66%
Total private pass. transport cost as % of metro. GRP	%	11.24%	12.39%	12.87%	6.75%	3.81%	11.52%	11.19%	13.11%	13.50%

Total public pass. transport cost as % of metro. GRP	%	0.55%	1.08%	0.85%	1.55%	1.60%	2.08%	2.44%	2.34%	4.16%
Traffic intensity indicators										
Private passenger vehicles per km of road	units/km	98.7	73.1	105.8	181.9	118.1	290.4	169.3	137.5	139.7
Pass. vehicles per km of road	units/km	98.9	73.3	106.1	183.1	121.7	300.4	184.4	138.9	154.5
Average road network speed	km/h	49.3	44.2	44.5	32.9	31.3	20.9	20.5	35.9	30.4
Transport energy indicators										
Private passenger transport energy use per capita	MJ/person	60,034	29,610	32,519	15,675	9556	10,555	2376	10,569	4052
Public transport energy use per capita	MJ/person	809	795	1044	1118	1500	1583	607	1012	1696
Energy use per private passenger kilometre	MJ/p.km	3.25	2.56	3.79	2.49	2.42	2.03	1.63	2.39	2.10
Energy use per public transport passenger kilometre	MJ/p.km	2.13	0.92	1.14	0.83	0.44	0.74	0.46	0.53	0.69
Air pollution indicators										
Total emissions per capita (CO, SO_2, VHC, NO_x)	kg/person	264.6	188.9	178.9	98.3	31.3	97.2	69.1	157.5	81.8
Total emissions per urban hectare	kg/ha	3563	2749	4588	5304	3894	12 952	13 357	7236	9211
Emissions per kilometre of motorized vehicle travel	kg/p.km	0.020	0.025	0.027	0.021	0.012	0.026	0.069	0.052	0.071
Transport fatalities indicators										
Total transport deaths per 100,000 people		12.7	8.6	6.5	7.1	5.9	20.7	10.4	18.3	13.2
Total transport deaths per billion passenger kilometres		7.0	6.8	7.1	9.6	7.4	29.2	37.4	29.3	34.0

Source: Data presented in Kenworthy and Laube (2001), *passim*.

persons per hectare) are higher than any of the other groupings, with the Low Income Other (LIO) category coming next with 122 persons per hectare (pph). A number of Asian cities have extremely high urban densities, a fact that has important implications for their range of transport-related options (Barter, 2000). Of the 13 cities with densities of more than 120 persons per hectare, 11 are Asian. They are Ho Chi Minh City (356 pph), Mumbai (337 pph), Hong Kong (320 pph), Seoul (230 pph), Taipei (230 pph), Manila (206 pph), Shanghai (196 pph), Jakarta (173 pph), Bangkok (139 pph), Chennai (133 pph) and Beijing (123 pph). The only other cities in the sample with densities that are comparable to the very dense Asian cities are Cairo (with 272 pph) and Barcelona (197 pph). Several other Asian cities have densities between 85 and 120 pph, which is still higher than any of the European cities except Barcelona. They are Guangzhou (119 pph), Osaka (98 pph), Singapore (94 pph) and Tokyo (89 pph). Only two of the Asian cities have urban densities that cannot be considered high. Kuala Lumpur has the lowest density of the Asian cities, with 58 pph, which is close to the average for the western European cities, and which might be characterized as middle-density. Sapporo in Japan (with 72 pph) also falls within the range of densities for European cities.

Wealth

The income per person of each urban region can also be important for transport development, for example by influencing which options are affordable. There are Asian cities at both extremes of income and at many levels in between. For example, the Asian groups of cities include one of the richest, Tokyo, with a gross regional product (GRP) per capita of US$45,425 and the poorest, Chennai, with only $396 per capita.[3] Asian cities offer some surprises when we compare their transport patterns and levels of wealth. We will see in subsequent sections that the wealthy Asian cities typically have much lower automobile dependence than cities with similar incomes per capita and even than many cities with much lower incomes. For example, private vehicle use in the HIA group tends to be comparable with or lower than even the Middle Income Asian (MIA) and Middle Income Other (MIO) groups, despite having more than three times the average GRP per capita of these groups.

Key policy-related contrasts in the sample of cities

This section focuses on policy-related themes arising particularly from Asian transport experiences. It examines data to illustrate contrasts in policy and practice among the cities in the sample, particularly the Asian cities. The focus of the section is on a pivotal choice: which modes of transport should be emphasized and deserve most policy attention? Two key dimensions of this choice are discussed, namely investment priorities and policy towards managing the pace of rising private vehicle use and motorization.

Transport investment priorities

Let us examine what the data set can tell us about recent and past investment priorities. First, we can gain some insight on past investment priorities by looking at their legacy in terms of the stock of transport infrastructure.

Starting with road provision, note that generous expressway provision in particular is a common hallmark of cities that place a high priority on private transport in transport policy (Thomson, 1977). Expressway provision per person is comparatively low in the HIA cities compared to other high income regions (0.022 metres per person, with Singapore being the exception having 0.044 metres per person). Western European cities have almost four times more expressway per person on average and US cities have seven times more than the HIA group, while all three regions have similar averages for GRP per capita. The lowest income groups (LIA and LIO) have low levels of expressway provision primarily due to an inability to afford such large investments. However, the MIA and MIO averages are higher than the HIA group on this indicator of commitment to private transport. Middle-income Kuala Lumpur stands out among the Asian cities for its particularly high level of expressway length per person (0.068 metres per person). It has been engaged in something of a frenzy of privatised toll-road building since the late 1980s. Decisions to build or not to build expressways are policy-driven and not merely an outcome of growing incomes *per se*.

The existence of intensely used reserved public transport routes is an indicator of a commitment to quality public transport. As motorisation progresses and there are fewer 'captive' riders of public transport, the need for investment in enhancements such as protected, higher speed rights-of-way (rail systems and busways) becomes more important in order to retain competitive speeds. The speed of the public transport system and its ratio to the speed of private transport provide simple indicators of success in giving priority to public transport. The HIA cities have respectable public transport operating speeds reflecting past investments in rail systems that now carry substantial proportions (30 per cent or more) of public transport passenger kilometres. Tokyo and Osaka stand out with very high public transport speeds (41 km/h and 50 km/h) that are much faster than private speeds (26 km/h and 33 km/h). Hong Kong and Singapore (and Seoul from the MIA group) have built substantial rail mass transit systems since the 1970s but retain an important role for buses in mixed traffic. Hence their public transport speeds are more modest (about 24 km/h) and slightly slower than private speeds. Nevertheless, they are doing much better than Taipei and Bangkok from the MIA group, which manage public transport speeds of only 13 km/h and 10 km/h respectively, or Kuala Lumpur where the public transport speed is 19 km/h, far slower than the private speed of 28 km/h. The LIA cities also have generally slow public transport, with speeds averaging only 17 km/h. The predominantly road-based public transport of most MIA and LIA cities suffers from the impacts of congestion.

The HIA group has the highest level of public transport seat kilometres per capita of all the groups (at over 5,000 seat kilometres per capita, more than 30 per cent higher than the Western European cities, the next most well-served group). Levels of public transport service are much lower in the middle and lower income Asian groups of cities (MIA and LIA), with less than half of the seat kilometres of service per capita of the HIA group and slightly less even than the levels in the other lower income groups, MIO and LIO.

Consistent with these facts, the HIA group also has the highest public transport use. The group has an average of 464 annual public transport boardings per capita and all the HIA cities, except Sapporo, have more than 45 per cent of the motorised transport task (passenger kilometres) on public transport. These figures are much higher than the MIA group with 274 boardings and 27 per cent of passenger kilometres on public transport (although Seoul and Bangkok do better than Kuala Lumpur or Taipei). The HIA public transport use is also higher than the nearest rival wealthy group of cities in Western Europe which has 297 boardings and 19 per cent. The Low Income Asian (LIA) and the Low Income Other (LIO) groups average high shares (51 per cent and 54 per cent) of motorised passenger kilometres on public transport but these high shares are in the context of much lower overall motorized mobility and are achieved with rather modest public transport usage (as measured for example by boardings). The Middle Income Other (MIO) group achieves the second highest usage of public transport on average with about 340 boardings per capita and 37 per cent of motorised passenger kilometres.

These findings suggest a surprisingly low level of public transport service and use in the MIA group, especially by Kuala Lumpur and Taipei, when compared with other relevant groups. Motorcycles in particular compete strongly with public transport in these cities, offering competitive speeds and relatively cheap mobility.

Table 15.2 shows data on levels of investment in roads and in public transport systems. These data are five-year averages of all investment from all sources in roads and public transport (new construction and maintenance). All regions have higher average investment in roads than in public transport but the imbalance between them is lowest in the HIA, WEU, LIO, and MIA groups. The high public transport investment levels in the MIA cities reflect a large, if belated effort to catch up, with large rail investments especially in Taipei but also in Kuala Lumpur and Bangkok during the 1990s, as well as significant expansion of Seoul's mass transit network. Taipei (and also Singapore) was in fact among the few cities in the sample with higher public transport investment than road. By contrast, the three Japanese cities seem to have been compensating for their earlier very high emphasis on urban rail by having among the highest levels of road investment per capita in the 1990s (see Chapter 9 of this volume).

The imbalance between investment in roads versus public transport is greatest in the LIA, USA and Canadian groups, each with well over 3 times more investment in roads than public transport. This imbalance in the LIA cities is particularly worrying since transport spending priorities in the early phases of motorization are likely to have a great influence on whether it is private transport or more balanced transport patterns that become firmly entrenched in the urban fabric. Of particular concern may be Guangzhou and Manila, which seem to have been investing about five to seven times more heavily in roads than public transport. In the light of the observations about local pollution dangers in Chinese cities in Chapter 11 (this volume) this imbalance in Guangzhou must obviously be of great concern.

Data on investment in facilities for walking and non-motorised vehicles is scarce and there is not the scope to examine this issue here. Non-motorised transport, particularly bicycle use, is especially vulnerable and easily discouraged by hostile street conditions. Efforts to provide a more welcoming environment for non-motorised modes clash head on with the demands of private vehicles, especially in dense urban environments where space is at a premium. Promotion of high roles for walking and cycling appears to have been very successful only in contexts where private motor vehicles have been restrained (by low incomes or by policy), for example in China and Vietnam in the 1980s, in Japanese cities, and in a number of northern European cities. Furthermore, success with promoting non-motorized transport is often an important complement to a strong role for public transport.

Managing the pace of motorization and private transport demand

This brings us to the second key area which reflects as well as shapes transport policy priorities. As incomes rise, cities face difficult choices over the pace of motorisation and the management of demand for private vehicle travel. Since the 1970s, many have argued that restraining private vehicles (by slowing the growth of their numbers and/or their use) is necessary in large, rapidly growing, rapidly motorising cities (Linn, 1983; Tanaboriboon, 1992, Chapters 10 and 14 of this volume). However, proposals for such measures always generate heated public debate.

We will argue that the experience from a number of the high-income and middle-income Asian cities strengthens the case for private vehicle restraint, especially at an early stage in the process when slowing the pace of motorisation can be the main restraint policy, as a crucial intervention in helping to create balanced and effective transport systems (Barter, 1999). There is also a strong link with the investment priorities discussed in the previous section, since restraining private vehicles apparently makes it much easier for a city to devote high priority to public transport over private transport.

Data from Table 15.2 on motorization and private vehicle use and Figure 15.1 (which shows private vehicle use versus income per capita) show that the HIA cities, and to a lesser extent the WEU cities, have levels of private car ownership and use that are surprisingly low considering their

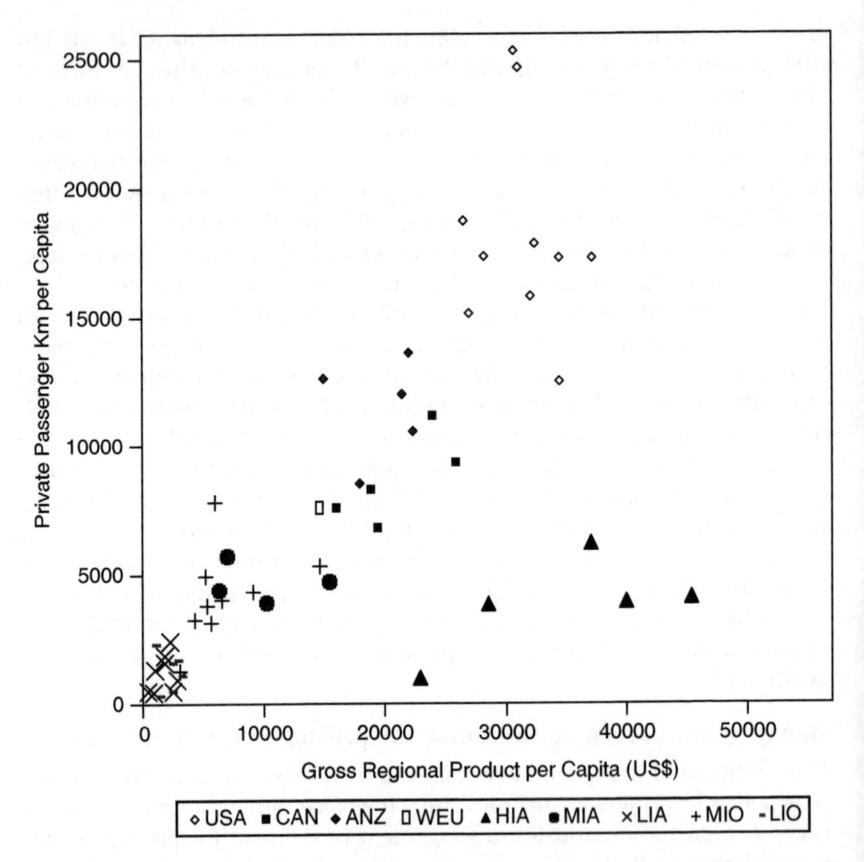

Figure 15.1 Private vehicle use versus income per capita in an international sample of cities, ca. 1995.

Source: Kenworthy and Laube (2001).

income levels. For example, the HIA cities on average have remarkably low private car ownership, with levels (217 cars per 1,000 people) that are comparable to the MIA cities (198 per 1,000) and lower than the average for the MIO cities (265 cars per 1,000). The USA and ANZ groups contrast with the Western European (WEU) group in having significantly higher motorization and vehicle use despite similar (or lower) average income levels per capita than the WEU group. The LIA cities have very low car ownership (of only 38 cars per 1,000) and use, as might be expected given their low-income levels.

The patterns for motorcycles are also striking, with the Asian groups standing out. The MIA cities tend to have the highest levels of ownership (154 motorcycles per 1,000 people on average). The average levels of motorcycle ownership in the HIA, MIA and LIA groups respectively are about two times, almost five times and about three times the average level of the next highest region, the Western European group of cities. The high motorcycle ownership in the middle and low income Asian groups (MIA and LIA) contrasts with the other non-Western groups of cities in the same income ranges (MIO and LIO), which both have very low motorcycle ownership of only 15 motorcycles per 1,000 people.

These contrasts can be further emphasized by examining specific cities from among the Asian groups. Hong Kong is most dramatic, with only 47 cars per 1,000 people and negligible motorcycle ownership despite a higher per capita income than many European, Australian and Canadian cities. Singapore is even wealthier but has constrained private passenger vehicle ownership to only 116 cars and 43 motorcycles per 1,000 people in our reference year. The three Japanese cities have much higher vehicle ownership (between 264 and 352 cars and between 45 and 138 motorcycles per 1,000 people) but these levels are still rather low considering these cities' very high levels of income per capita.

Among the Middle Income Asian (MIA) group of cities, Bangkok (with 249 cars and 205 motorcycles per 1,000 people) and Kuala Lumpur (with 209 cars and 175 motorcycles per 1,000 people) most clearly have higher motorisation than expected on the basis of income.[4] In this they are similar to many of the MIO cities which also tend to have high car ownership relative to incomes. Seoul and Taipei show some modest restraint of private vehicles according to the data and, despite being richer than Bangkok or Kuala Lumpur, they fall between them and the HIA cities in terms of motorisation relative to income, with 160 and 175 cars and 39 and 197 motorcycles per 1,000 respectively.

What circumstances and policy decisions underpin these numbers? First let us mention Kuala Lumpur and Bangkok, where the only restraint has been modest price disincentives related to tariff protection for local motor industries. These have had a decreasing impact as incomes have risen, as local production of low-priced vehicles has increased, and as protection has been lowered (at least in Thailand). Despite acute traffic problems, particularly in Bangkok's denser urban fabric, neither city has proposed any policy measures to slow down the pace of motorisation. Usage disincentives have never been pursued seriously (Barter, 1999).

At the other extreme within the Asian groups of cities, low vehicle ownership and use in Hong Kong and Singapore can be directly attributed to their well-documented restraint policies, especially to contain ownership since the early 1970s (Ang, 1996; Wang and Yeh, 1993). In Hong Kong's case, this began with car ownership levels of less than 30 cars per 1,000 persons. Singapore's restraint began when car ownership was a little higher at about 70 per 1,000 persons plus about 50 motorcycles per 1,000 persons (Barter, 1999).

The Japanese cities and Seoul and Taipei are now between the above two extremes in terms of their restraint of private vehicles. Seoul and the

Japanese cities contrast in many ways but their histories of private vehicle restraint have much in common (Barter, 1999). Both Japan (prior to the 1960s) and Korea (prior to the 1980s) had macroeconomic strategies which involved severe restraint on private consumption, including the purchase of private vehicles. Therefore their motor vehicle ownership remained extremely low (only 16 cars per 1,000 people in Tokyo in 1960 and the same in Seoul in 1980) despite each country already enjoying considerable economic success (Barter, 1999). Both Japan and Korea subsequently relaxed their restraints on vehicle ownership and allowed a burst of motorisation (in the 1960s in Japan and since the mid-1980s in Korea). However, some disincentives to vehicle ownership and usage remain in force or have been introduced as congestion has become a greater problem (Barter, 1999). In addition, there are important legacies of the earlier period of restraint, such as extensive and well-used rail systems and considerable transit-oriented development in both places.

Taipei's experience has much in common with those of Bangkok and Kuala Lumpur, including very high levels of motorcycle ownership and use. Like these two cities, Taipei also faced rather extreme traffic congestion problems by the mid 1990s and a low level of public transport use. However, Taipei's motorisation was somewhat slower than Kuala Lumpur's or Bangkok's. It has currently reached a similar level of motorisation but with a very much higher level of income. This difference may relate indirectly to the need to import oil and to the lack of a significant car manufacturing industry in Taiwan, although the motorcycle/scooter industry is among the largest in the world. A very high-density urban fabric may also be important by making traffic impacts such as congestion, pollution and parking shortages emerge very quickly as motorisation took off. It seems that for various reasons and without much fanfare, the costs of buying and using private cars in Taiwan have been kept at a relatively high level. Data from this study suggest that in the mid-1990s overall costs per passenger kilometre for private vehicles were comparable with Japanese and Swiss cities. The World Bank also reports that in 1994 the price of gasoline in Taiwan was about double that in Thailand (World Bank, 1996). It remains to be seen if these subtle differences between Taipei and the Southeast Asian cities will be important for subsequent trends. High motorcycle use and rising car ownership had stifled Taipei's public transport development up until the mid 1990s, but there are signs recently of improvements for public transport via a very successful bus priority system, rapidly expanding mass transit and moves to tighten regulation of parking, including parking of motorcycles (Her, 2001; Hwang, 2001).

Benefits of slowed motorization for public transport development

In a number of the cities examined above, deliberately slowing down the motorisation process was an important factor in allowing public transport to build its role, even as incomes increased (Barter, 1999). Examples include the Japanese cities, Singapore and Hong Kong. For a time, Seoul was also an example of this phenomenon. Many western European cities also probably offer support for this argument, especially when they are compared with the other western groups of cities in this sample, although it is not possible to investigate this question here.

Tokyo, Osaka, Hong Kong, Singapore and Seoul all share a history of having strongly curtailed motorisation for a significant period at an early stage (well before motorization reached around 150 vehicles per 1,000 people). In the cases of Hong Kong, Singapore and Seoul, high-quality mass transit systems were not yet in operation at the time that private vehicle restraint began. In fact, slow motorization despite rapidly rising incomes allowed all of these cities a window of opportunity during which they could continue to invest in public transport and eventually provide substantial, high-quality public transport systems *before* private vehicle ownership reached 150 vehicles per 1,000 persons. In doing so these cities avoided many of the transport-related problems and pitfalls that befall many cities with rapidly rising incomes. They were able to maintain bus-based public transport usage at a high level until mass transit became affordable and was built. Public transport never became the mode of last resort, or to be seen as only for the poor, in these cities. Seoul and the Japanese cities, where constraints on motorisation have been relaxed, provide some evidence that the earlier policies helped to provide an irreversible legacy. Tokyo, Osaka and Seoul now have a 'critical mass' of traffic-segregated public transport, which unlike buses in mixed traffic, will not enter a vicious downward cycle even if the roads become grid-locked. Public transport in these cities is therefore likely to be seen as part of the solution.

These experiences contrast with those of Kuala Lumpur, Taipei and Bangkok where motorisation was able to reach rather high levels before substantial mass transit systems were in place and public transport use became stigmatised as a mode of last resort (Barter, 1999). In Taipei, and especially in Kuala Lumpur, public transport's mode share has dropped to low levels, which will be difficult to reverse even with expanding mass transit. Bangkok retains surprisingly high public transport ridership (primarily captive riders of the very slow bus system), but public transport improvements are very slow and ridership seems likely to be extremely vulnerable in the event of a resumption of economic growth and further motorization. Some of the LIA cities seem likely to follow this unhappy trend unless serious steps are taken to slow motorisation in order to buy time to enhance the alternatives to private vehicles. Ho Chi Minh City already has a very minimal role for public transport and is dominated by motorcycles.

This argument about the importance of slowing the pace of motorization helps to place mass transit investments into a new perspective. Most commentators agree that expensive investments in rail mass transit are extremely difficult for low-income cities (Fouracre *et al.*, 1990). The argument here suggests that the deliberate slowing of motorization can play a key role in allowing a city to set out on a transit-oriented development path. Such a strategy prevents rapid motorization from undermining the role of public transport and instead allows the role of public transport to build up gradually in conditions of rising incomes but low vehicle numbers and relatively low pressure to invest heavily in roads. The delay in motorization means that the decision to spend on mass transit can be delayed until it is relatively affordable for the city and yet still be guaranteed high ridership. This is in contrast to the consequences of allowing unrestrained motorization with congestion and modal competition causing bus services and usage to deteriorate, creating intense pressure to expand roads, and making investments in mass transit appear less and less viable.

Prospects for restraint in low-income cities

The comments above suggest that the early years of the new millennium will be a crucial time for the lower-income groups of cities, especially those enjoying a measure of economic success. Efforts to slow motorization could help buy time to build a more balanced transport system and avoid some of the worst impacts of explosive motorization. It is generally too soon to tell whether any of these cities are likely to do this but the signs are mostly not promising. Glimmers of hope do appear from time to time, such as Bogotá's referendum in October 2000 which committed the city to aiming for a car-free system by 2015.

Ho Chi Minh City is unique in this sample of cities (but perhaps representative of a small group of other cities in Indochina) in experiencing extremely rapid growth in motor cycle ownership which has almost destroyed what little public transport there had been and is in the process of replacing bicycles. With 291 motorcycles per 1,000 people Ho Chi Minh City has by far the highest motorcycle ownership in this sample of cities. Only Bangkok (with 205 per 1,000 people), Taipei (with 197), Kuala Lumpur (175) and Jakarta (168) have anything like this many motorcycles per capita. There are as yet no signs of efforts to restrain this motorcycle-based motorisation process (MVA, 1997). However, car ownership is still currently very low and, in addition to low income levels, this may be partly attributed to relatively high ownership fees (Heil and Pargal, 1998). The city's motorcycle dominated situation is unprecedented, but the experiences of Bangkok and Taipei suggest that, combined with very high urban densities, the impact of motorcycles on public transport sets the scene for extreme traffic-related problems when economic growth later brings in more private cars.

Jakarta and Manila have both faced rapid motorisation in the 1990s which has been slowed or halted since the East Asian economic crisis that began in 1997. Neither has taken steps to slow motorization but both have begun attempts to restrain the use of private vehicles in the most congested places (Barter, 1999). Although modest, these efforts have been accompanied by intense debate, which is likely to intensify if the economic situation improves and rapid motorization resumes. Indonesia's efforts to reduce the huge fuel subsidy has repeatedly met with fierce political resistance. Neither country (or city) is considering efforts to slow motorization itself (Barter, 1999).

Chinese cities are in the early stages of a boom in car ownership. National policies increasingly aim to increase private car ownership in order to encourage the motor vehicle industry. Moves to reduce national-level taxes and fees on cars will soon be complemented by reduced tariffs as China enters the World Trade Organization (*China Daily*, 22 April 2001). Beijing shows no sign of restricting cars significantly but some cities, most notably Shanghai, have been deliberately slowing the growth of vehicle ownership by various means. Both car and motorcycle ownership are controlled through an auction of certificates to register. These restrictions are despite the status of the Shanghai area as the most important focus of the Chinese motor industry. Shanghai has much lower car ownership than Beijing despite its higher income levels. However, there are signs that the severity of these restrictions is gradually being eased (*Shanghai Star*, 28 January 2000).

Indian cities are also in the early stages of a boom in private vehicle ownership, with both car and motorcycle numbers rising steeply in the wake of deregulation of the vehicle industry and vehicle imports along with rising numbers of middle income urban dwellers (see Chapter 10 of this volume). There are few, if any, signs of efforts to restrain the rate of motorization, although public debate over the impacts of vehicles is increasing. The national government is also gradually trying to reduce the burden of fuel subsidies on the national budget, which will probably lead to slightly higher gasoline prices.

Transport impacts

Let us now briefly consider data on the impacts of transport. These suggest that policies seeking a balance in transport priorities and avoiding dominance by private vehicles are important for minimizing the negative impacts of transport.

Cost-effectiveness

First, the data set offers information on cost effectiveness suggesting that an emphasis on the alternatives to private vehicles offers the most thrifty strategy when viewed at a city-wide level. One measure of the cost-effectiveness of urban transport is how much of a city's GRP has to be spent moving people around (public and private transport operating and investment costs). It is obviously desirable to minimize this expenditure while at the same time providing high (or at least satisfactory) levels of access to services, goods and human opportunities.

If we assume that all high-income cities come close to providing satisfactory access for their residents then it is striking that on average the least car-oriented groups are able to do so much more economically than the car-dominated wealthy groups. The HIA group expends on average only 5.4 per cent of GRP (3.8 per cent on private transport and 1.6 per cent on public transport), ahead of WEU cities at 8.3 per cent and far more thriftily that the more car-dependent regions which spend between 11.8 and 13.7 per cent of GRP on passenger transport, the vast majority of which is private transport (11.2 to 12.9 per cent).

In the middle and low-income groups we cannot necessarily assume that access needs are being adequately met and levels of spending on the transport task in these groups of cities are typically relatively modest in absolute terms. Nevertheless, this spending amounts to high proportions of GRP (from 13.6 per cent in the MIA group to 17.7 in the LIO group). As we have seen above, there is a tendency among these groups to have comparatively high levels of motorization and vehicle use relative to their economic development level. An exception, is the group of three Chinese cities, which spend a slightly more modest 10.7 per cent of GRP on average on passenger transport. This is almost certainly due to their emphasis on cost-effective non-motorised transport, particularly bicycles.

The data on spending also demonstrate that when public transport is well used it is very cost-effective relative to its proportion of the motorized passenger task. This is most clearly shown by the lower income groups of cities. For example, in the LIA group, 51 per cent of motorized passenger kilometres are carried by public transport for expenditure of 2.4 per cent of GRP (or about 18 per cent of the transport spending), versus 49 per cent carried by private vehicles for 11.2 per cent of GRP (or about 82 per cent of passenger transport spending)! By contrast the USA group has poorly patronized public transport which serves less than 3.0 per cent of the passenger transport task on average but which accounts for 4.7 per cent of the transport spending. But note that the Canadian group's well-used public transport does better, serving almost 10 per cent of passenger kilometres but taking only 6.2 per cent of the transport spending.

Energy use and air pollution

It is also desirable for a transport system to minimize the energy use and air pollution emissions attributable to transport. Again the HIA cities do well. HIA cities are by far the least energy intensive of the high income groups of cities, having a per capita consumption (private and public transport energy) that averages only 18 per cent of the US group average and having the most energy-efficient private and public transport systems per passenger kilometre. The HIA transport energy usages per capita are in fact slightly lower than those of both middle income groups in Table 15.2.

HIA cities, which combine low car use with strong emissions regulations for vehicles, have dramatically lower per capita emissions of CO, SO_2, VHC and NO_x than any other group of cities, including the middle and low income groups. However, HIA cities also tend to have high urban densities. This results in a high spatial intensity of emissions, higher on average than those of the USA and ANZ groups, despite the much higher emissions per capita of those groups. Similarly, despite modest emissions per capita the highest density groups of cities, the MIA and LIA groups, have the highest levels for emission per hectare. This highlights the need for dense cities to work especially hard to ensure low per capita emissions. Unfortunately, currently MIA and LIA cities tend to have a nasty combination of factors with dense urban forms, rising use of motor vehicles, and poor emissions control.

Safety

Contrasts in transport safety outcomes also reveal benefits from restraining private vehicle use. The Middle Income Asian (MIA) group has the highest transport death rate per 100,000 people of all of the groups (at 20.7, slightly higher than the MIO group with 18.3), while the more transit-oriented HIA group has the lowest (at 5.9). The problem for the middle-income groups of cities is that they combine both very high rates of transport deaths per billion passenger kilometres with relatively high vehicle use. The two low-income groups (LIA and LIO) have even higher deaths per billion passenger kilometres but because of lower vehicle use this translates to more moderate rates of deaths per 100,000 people. The USA cities also display the safety problems of high vehicle use. Even though the USA group has among the lowest deaths per billion passenger kilometres, their extremely high levels of vehicle use mean that they still have the highest rate of transport deaths per 100,000 people from among the five high-income groups of cities (at 12.7 versus 8.6, 6.5, 7.1 and 5.9 for the ANZ, CAN, WEU and HIA groups respectively).

Conclusion

Evidence from cities in Asia provides important policy insights for low-income cities everywhere, especially those that are beginning to enjoy some economic success. The evidence in this chapter suggests that an early decision to prioritize public transport and non-motorized transport investments over private transport-oriented investments can bring important long term benefits. Such investment priorities are made enormously easier to carry out in developing cities if the pace of motorisation can be deliberately slowed down, especially during the early stages of the process which often accompany periods of rapid economic growth and urbanization. The need for better balanced transport systems is particularly acute for cities that are already large and dense, as is very common in cities throughout the low-income parts of the world, since dense cities are particularly vulnerable to the negative impacts of traffic.

Notes

1. References to 'Asian cities' in this chapter refer to these three sub-regions and do not include cities in other Asian sub-regions, such as Teheran, Riyadh or Tel Aviv in Southwest Asia.
2. Urban density is calculated by dividing the population of the metropolitan area by its urbanized area. In order to ensure comparability, the urbanized area is defined independently of national definitions ('urbanized' or 'built up' area) which may be inconsistent. The urbanized area is the sum of the areas taken up by the following land-uses including: residential, industrial, offices, commercial, public utilities, hospitals, schools, cultural uses, sports grounds, wasteland (urban), transport facilities, and small parks and gardens. This urbanized area is based on the best available land-use inventory data at an appropriate level of resolution or, in some cases, based on planimeter measurements from maps or aerial photographs.
3. Gross Regional Product (GRP) is a measure of the income of a sub-national region. GRP data is often available, or can be calculated, for functional economic regions through national statistical agencies.
4. Motorization levels in Bangkok and Kuala Lumpur are clearly high, despite some concern that these figures may be slightly inflated due to inaccurate official data, particularly in the case of Bangkok.

Bibliography

Adriaanse, A. *et al.*, *Resource Flows: the Material Basis of Industrial Economies* (Washington, DC: World Resource Institute, 1997).

American Methanol Institute, *Beyond the Internal Combustion Engine: the Promise of Methanol Fuel Cell Vehicles* (Report prepared by Breakthrough Technologies Institute for the American Methanol Institute) (Washington, DC: American Methanol Institute, 2001).

Anderson, W.K., *Roads for the People* (South Melbourne: Hyland House, 1994).

Ang, B.W., 'Urban transportation management and energy savings: the case of Singapore', *International Journal of Vehicle Design*, 17/1: 1–12 (1996).

Anshelm, J., *Socialdemokraterna och miljöfrågan: en studie i framstegstankens paradoxer* [*The Social Democrats and the Issue of Environment: a Study of the Paradoxes of the Idea of Progress*] (Stockholm: Symposion, 1995).

Appleyard, D., *Livable Streets* (Berkeley and Los Angeles: University of California Press, 1981).

Applin, G., Beggs, P., Brierley, G., Cleugh, H., Vurson, P., Mitchell, P., Pitman, A. and Rich, D., *Global Environmental Crises: an Australian perspective*, 2nd edn (Melbourne, Australia: Oxford University Press, 1999).

APTA, *Attitudes to Rail Transit* (http//:www.apta.com, 2000).

Aschauer, D.A., 'Is public expenditure productive?', *Journal of Monetary Economics*, 23: 177–200 (1989).

Aschauer, D.A., 'Why is infrastructure important?', in Munnell, A.H. (ed.), *Is There a Shortfall in Public Capital Investment?* (Conference Series No. 34, Boston: Federal Reserve Bank of Boston, 1990).

Ashton, P. *The Accidental City: Planning Sydney Since 1788* (Sydney: Hale and Iremonger, 1995).

Asian Development Bank (ADB), *Development of Long-Term Plan for Development of Expressways in India*, Report submitted to the Ministry of Surface Transport (Delhi: Government of India, Delhi Consulting Engineering Services, 1991).

Athanasiou, T., *Slow Reckoning: the Ecology of a Divided Planet* (London: Verso, 1998).

Axcess Australia, *The Car Immediate Power*, (http://www.axcessaustralia.com/car/power.asp, October 2000).

Baer, P., Harte, J., Haya, B., Herzog, A.V., Holdren, J., Hultman, N.E., Kammen, D.M., Norgaard, R.B. and Raymond, L., 'Equity and greenhouse gas responsibility', *Science*, 289: 22 (2000).

Baier, von Reinhold and Schaefer, K.H., 'Innenstadt-verkehr und Einzelhandel in der Städtetag: Zeitschrift für Kommunalen', *Praxis und Wissenschaft*, 8 (1997), 559–68.

Banister, D., 'Equity and acceptability questions in internalising the social costs of transport', in ECMT (European Conference of Ministers of Transport), *Internalising the Social Costs of Transport* (Paris: ECMT/OECD, 1994) pp. 153–73.

Barricade (Anonymous authors, Australian Independence Movement: Melbourne, 1978).

Barter, P.A., *'An international comparative perspective on urban transport and urban form in Pacific Asia: responses to the challenge of motorization in dense cities'*, PhD dissertation (Murdoch, Western Australia: Murdoch University, 1999).

Barter, P.A., 'Transport dilemmas in dense urban areas: examples from Eastern Asia', in Jenks, M. and Burgess, R. (eds), *Compact Cities: Sustainable Urban Forms for Developing Countries* (London: Spon, 2000) pp. 271–84.
Barthes, R., *Mythologies* (London: Jonathan Cape, 1972).
Barton, H., *Sustainable Communities: the Potential for Eco-Neighbourhoods* (London: Earthscan, 2000).
Beck, U., Giddens, A. and Lash, S., *Reflexive Modernization, Politics, Tradition and Aesthetics in the Modern Social Order* (Cambridge: Polity Press, 1994).
Beder, S., *Global Spin: the Corporate Assault on Environmentalism*, rev. edn (Melbourne: Scribe, 2000).
Beed, C., *Melbourne's Development and Planning* (Melbourne: Clewara Press, 1981).
Bernard, M., 'Ecology, political economy and the counter-movement: Karl Polanyi and the second great transformation', in S. Gill and J.H. Mittelman (eds), *Innovation and Transformation in International Studies* (Cambridge: Cambridge University Press, 1997) pp. 73–89.
Bernitz, U. and Kjellgren, A., *Europarättens grunder* (*The Foundations of European Law*) (Stockholm: Norstedts, 1999).
Birch, S., 'Ford's focus on the fuel cell', *Automotive Engineering International*, New York (June 2001): 25–28.
Björklund, P., Tjäder, C. and Wiberg, R., *Europeisk transportpolitik: En översikt* (*European Transport Politics, an Overview*) (Stockholm: Sveriges Transportindustriförbund 2000).
Black, W.R., 'Socio-economic barriers to sustainable transport', *Journal of Transport Geography*, 8 (2000): 141–7
Black, W.R., 'An unpopular essay on transportation', *Journal of Transport Geography*, 9 (2001): 1–11.
Blomen, L.J.M.L. and Mugerwa, M.N., *Fuel Cell Systems* (New York: Plenum Press, 1993).
Blowers, A., 'We can't go on as we are – the social impacts of trends in transportation' in Blessington, H. (ed.), *Urban Transport: Proceedings of the Institution of Civil Engineers conference Birmingham 9–10 March 1995* (London: Thomas Telford Publications, 1995) pp. 27–33.
Boardman, B., Fawcett, T., Griffin, H., Hinnells, M., Lane, K. and Palmer, J., *DECADE – Domestic Equipment and Carbon Dioxide Emissions – Two Million Tonnes of Carbon* (Oxford: University of Oxford Environmental Change Unit, 1997).
Böge, A., 'The well travelled yoghurt pot: lessons for new freight transport policies and regional production', *World Transport Policy and Practice*, 1/1 (1995): 7–11.
Bombay Metropolitan Regional Development Authority (BMRDA), *Draft Development Plan of Bombay Metropolitan Region* (Bombay: Bombay Metropolitan Regional Development Authority, 1973).
Boukhalfa, N., Goldman, A., Goldman, M. and Sigmond, R.S., *Proceedings of the International Symposium on Plasma Chemistry*, 2 (1987): 787–92.
Breheny, M., 'The contradictions of the compact city: a review', in M. Breheny (ed.), *Sustainable Development and Urban Form* (London: Pion, 1992) pp. 138–59.
Bressers, H.T.A. and Plettenburg, L.A., 'The Netherlands', in M. Jänicke and H. Weidner (eds), *National Environmental Policies: a Comparative Study of Capacity-Building* (Berlin: Springer Verlag, 1997) pp. 109–32.
Brög, W., 'Does anybody still walk nowadays?: Mobility facts from Germany', Perth, Western Australia, First International Walking Conference, February 2001.
Brown, L., 'Feeding nine billion', in L. Brown and C. Flavin *et al.* (eds), *State of the World 1999* (London: Earthscan, 1999) pp. 115–32.
Brundtland, G.H. (chair), *Our Common Future: A report from the World Commission on Environment and Development* (Oxford: Oxford University Press, 1987).

Bryant B. and Mohai, P. (eds), *Race and the Incidence of Environmental Hazards* (Boulder, CO: Westview Press, 1992).
Bullard, R., *Dumping in Dixie* (Boulder, CO: Westview Press, 1990).
Button, K.J., *The Economics of Urban Transport* (Faruborough, UK: Saxon House, 1977).
Button, K. and Verhoef, E.T., 'Transport at the edge of mobility and sustainability', in K. Button, P. Nijkamp and H. Priemus (eds), *Transport Networks in Europe: Concepts, Analysis and Policies* (Cheltenham: Edward Elgar, 1998) pp. 333–50.
Burgmann, M. and Burgmann, V., *Green Bans, Red Union: Environmental Activism and the New South Wales Builders Labourers' Federation* (Sydney: University of New South Wales Press, 1998).
Campbell C.J., *The Coming Oil Crisis* (Multi-Science Publishing Co. and Petroconsultants, 1997).
Campbell C.J., and Laherrère J.H., 'The end of cheap oil', *Scientific American* 278/3 (1998): 80–6.
Castells, M., *The Informational City: Information Technology, Economic Restructuring and the Urban Regional Process* (Oxford: Blackwell, 1989).
Castells, M. and Hall, P., *Technopoles of the World* (London: Routledge, 1994).
Cervero, R., 'Unlocking suburban gridlock', *American Planning Association Journal*, Autumn (1986): 389–406.
Chen, H.S., *Shanghai Urban Transport Analysis and Prediction* (Shanghai: Shanghai Science and Technology Press, 1998).
Chen, S.L., Lyon, R.K. and Seeker, W.R., 'Advanced Non-Catalytic Post Combustion NO_x Control', *Environmental Progress*, 10/3 (1991): 182–5.
China Daily (newspaper), 'Domestic Car Pricing Control Lifted', 22 May 2001. Online (<www.china.org.cn/english/13108.htm>July 2001).
Christiansen, P.M., 'Denmark', in P.M.Christiansen (ed.), *Governing the Environment: Politics, Policy, and Organization in the Nordic Countries*, Nordic Council of Ministers (Copenhagen: Nord, 1996) pp. 29–102.
COM 23, *Towards Sustainability: a European Community Programme of Policy and Action in Relation to the Environment and Sustainable Development* (1992).
COM 46, *Green Paper on the Impact of Transport on the Environment: a Community Strategy for 'Sustainable Mobility'* (1992).
COM 494, *The Future Development of the Common Transport Policy – a Global Approach to the Construction of a Community Framework for Sustainable Mobility* (1992).
COM 601, *European Commission Green Paper: the Citizens Network – fulfilling the Potential of Public Passenger Transport in Europe* (1995).
COM 689, *A Community Strategy to Reduce CO_2 Emissions from Passenger Cars and Improve Fuel Economy* (1995).
COM 88, *A Community Strategy to Combat Acidification* (1997).
COM 2000, *Agenda 2000: I) For a Stronger and Wider Union, II) The Challenge of Enlargement* (1997).
COM 431, *Developing the Citizens Network: Why Good Local and Regional Passenger Transport Is Important, and How the European Commission Is Helping To Bring It About* (1998).
COM 466, *Fair Payment for Infrastructure Use* (1998).
COM 716, *Sustainable Mobility: Perspectives for the Future* (1998).
COM/99/0640 final, *Air Transport and the Environment: Towards Meeting the Challenges of Sustainable Development* (1999).
COM 31, *Environment 2001: Our Future, Our Choice: the Sixth Environmental Action Programme of the European Community 2001–2010* (2001).
Commonwealth Government, *Portfolio Budget Statements 2000–01, Transport and Regional Services Portfolio* (Canberra: Government Printer, 2000) (available at http://www.dotrs.gov.au/budget/).

Country Roads Board, 'Hume Freeway, Eastern Freeway to St George's Road' Planning and design Review Committee Friday 14th February 1975, Notes of Discussion' (1974 mimeo in possession of N.Low, The Faculty of Architecture, The University of Melbourne, Vic 3010).

CTP, *The Common Transport Policy: Sustainable Mobility: Perspective for the Future: European Commission DG VII Transport* (Commission Communication to the Council, European Parlament, Economic and Social Committee and Committee of the Regions 1999).

Curtis, C., *Integrated Land Use and Transport Planning Policies: a Review of Selected Initiatives outside Australia and their Applicability to Strategic Land Use Planning in Perth* (Perth: Western Australian Planning Commission, 1998).

Curtis, C., *Future Perth: Perth Metropolitan Region: Transport*. Working Paper No. 7 (Perth: Western Australian Planning Commission, 2001).

Daly, H., *Beyond Growth: the Economics of Sustainable Development* (Boston: Beacon Press, 1996).

Danish Transport Council, *Dansk Transportpolitik – En Oversigt* [*Danish Transport Policy: an Overview*] (Copenhagen: The Danish Transport Council, 1993) report no. 93–01.

Davidson, K. 'PM finds a partner in policy pollution', *The Age* (Melbourne), (12 August 1999): 7.

Davidson, K., 'Transport spending is badly off track', *The Age*, (Melbourne) (15 May 2000): 15.

Davison, G., 'Dream Highways – Automobilising Melbourne 1945–1975', *Proceedings: Urban History Planning Conference 27–30 June, Volume 1*, (Canberra: Urban Research Program, 1995) 19 pages, unnumbered.

D'Monte, D. 'Mumbai's urban transport infrastructure: the missing links', paper presented at the 'National Conference on Globalisation and Environment' (Mumbai: SNDT University, 2001).

Denmark, Ministry of Transport, *The Danish Government's White Paper on Transport and the Traffic Plan 'Traffic 2005'*, Summary in English (Copenhagen: Ministry of Transport, 1994).

Department of Planning and Urban Development *Perth: Metroplan* (Perth: DPUD, 1990).

Department of Transport, *Metropolitan Transport Strategy 1995–2029* (Perth: Government of Western Australia, 1995).

Department of Transport, *Better Public Transport: Ten-Year Plan for Transperth 1998–2007* (Perth: Government of Western Australia, 1998).

Diesendorf, M., 'All choked up over our love affair with the car', *Sydney Morning Herald* (30 November 1999): 19.

Dillard, D., *The Economics of John Maynard Keynes: the Theory of a Monetary Economy* (London: Crosby Lockwood, 1948).

Dingle, T. and Rasmussen, C., *Vital Connections: Melbourne and Its Board of Works* (Ringwood, Australia: McPhee Gribble/Penguin, 1991).

Dobinson, K., 'New public transport vision a must to stall our dependence on cars', *Sydney Morning Herald* (26 February, 2002): 15.

Dobson-Mouawad, D., Dobson-Mouawad, L. and Pearce, D., *Air Pollution and Health* (London: British Lung Foundation, 1998).

Eckersley, R., *Environmentalism and Political Theory: Towards an Ecocentric Approach* (London: UCL Press 1992).

EFTE (European Federation for Transport and the Environment), *Briefing Paper on Road Pricing* (EFTE: Brussels, 1996).

Elkington, J., *Cannibals with Forks: the Triple Bottom Line of 21st Century Business* (New Society Publications, 1998).

Environment Protection Authority (EPA), *Estimation and Evaluation of Cancer Risks Attributed to Air Pollution in SW Chicago,* Final summary report, EPA Region 5, Chicago (Washington DC, USA: Environmental Protection Agency, 1993).

European Environment Agency (EEA), *Are We Moving in the Right Direction?: Indicators on Transport and Environment Integration in the EU,* Environmental Issues Series No. 12 (Copenhagen: European Environment Agency, 2000).

EC (European Commission) *1992: the Environmental Dimension,* Task Force Report on the Environment and the Internal Market (Brussels: European Commission, undated).

EC (European Commission) *Growth, Competitiveness and Employment* (Brussels: European Commission, 1993).

EC (European Commission), *Transport: Sustainable Mobility* (http://www.europa.eu.int/comm/tra ... mes/mobility/english/sm_4_en.html) accessed: 30/10/2001.

EC (European Commission), *The Citizen's Network: Fulfilling the Potential of Public Passenger Transport in Europe* (Brussels: European Commission, 1996a).

EC (European Commission), *European Sustainable Cities Report* (Brussels: European Commission, Expert Group on the Urban Environment, 1996b).

EC (European Commission), *Improving Energy Efficiency and Reducing Gas Emissions in Urban Transport* (Brussels: European Commission, Directorate General XVII Energy, 1997).

EC (European Commission) European Transport Policy for 2010: *Time to Decide,* Transport White Paper (Brussels: European Commission, 2001).

Economist, The, 'California's Power Crisis', (20 January 2001): 57–9.

European Union Emission Standards, 'Diesel cars and light-duty trucks', (Luxembourg: European Union, 1998).

Fouracre, P.R., Allport, R.J. and Thomson, J.M., 'The performance and impact of rail mass transit in developing Cities', *TRRL Research Report 278*. (Crowthorne, UK: Transport and Road Research Laboratory, Overseas Unit, 1990).

Freeman, C. (ed.), *The Longwave in the World Economy* (Aldershot, International Library of Critical Writings in Economics, Edward Elgar, 1996).

Freeman, C. and Perez, C., 'Structural crises of adjustment, business cycles and investment behaviour' in G. Dosi, D. Freeman, R. Nelson, G. Silverburg and L. Soete (eds), *Technical Change and Economic Theory* (London: Pinter, 1988).

Freeman, C. and Soete, L., *The Economics of Industrial Innovation,* 3rd edn (London: Frances Pinter, 1997).

Freund, P. and Martin, G. *The Ecology of the Automobile* (Montreal: Black Rose Press, 1993).

Fulton, L., *Statistical Effects of Induced Travel in the US Mid-Atlantic Region,* Presentation to the Transportation Research Board (May 2001).

Fuji, S. and Kitamura, R., 'Evaluation of trip-inducing effects of new freeways using a structural equations model system of commuters' time use and travel', *Transportation Research B,* 34 (2000): 339–54.

Giddens, A., *Modernity and Self-Identity: Self and Society in the Late Modern Age* (Cambridge: Polity Press, 1991).

Girouard, M. *Cities and People* (New Haven, Conn. and London: Yale University Press, 1985).

Gold T., 'The deep hot biosphere', *Geojournal Library,* 9 (1999): 85–92.

Goodwin, P.B., 'Extra traffic induced by road construction: empirical evidence, economic effects and policy implications', *European Conference of Ministers of Transport, Round Table,* (1997): 140.

Gordon, D., *Steering a New Course: Transportation, Energy and the Environment* (Covelo: Island Press, 1991).

Gotz, K. *et al.*, *'Citymobil', Mobilitätsstile – Ein Sozial-Ökologischer Untersuchungsansatz* (Frankfurt, 1997).

Gramlich, E., 'Infrastructure investment, a review essay', *Journal of Economic Literature*, 32: (1994): 1176–96.

Habermas, J., *Theorie des kommunikativen Handelns* (*The Theory of Communicative Action*) vols I–II (Frankfurt am Main: Suhrkamp, 1981).

Hajer, M., *The Politics of Environmental Discourse, Ecological Modernisation and the Policy Process* (Oxford: Oxford University Press, 1995).

Hall, P., 'Reflections past and future in planning cities', *Australian Planner*, 34/2 (1997): 83–9.

Hall, P.A., *The Innovative City* (Melbourne, Australia: Proceedings of OECD Conference, 1994).

Hall, P.A., 'European perspective on the spatial links between land use, development and transport', in D. Banister (ed.), *Transport and Urban Development* (London: Spon, 1995): 65–88.

Hall, P.A. and Pfeiffer, U., *Urban Future 21: a Global Agenda for Twenty-First Century Cities* (London: Spon, 2000).

Hammond, B., 'Buses, bicycles and small town revivals', in *US–European Perspectives on the Climate Change Debate* (Washington, DC: Center for Clean Air Policy, 1994) pp. 33–41.

Hansen, M., 'Do highways generate traffic?', *Access*, 7 (1995).

Harada, K. and Aoki, E., *Nippon no Tetsudo: 100 Nen no Ayumi Kara* (*The Railways of Japan: from Its 100 Years of History*) (Tokyo: Sanseido, 1973).

Hart, T., 'Transport and the city' in R. Paddison (ed.), *Handbook of Urban Studies* (London and New York: Sage, 2001) pp. 102–23.

Harvey, D., *Social Justice and the City* (London: Edward Arnold, 1973).

Hass Klau, C., 'Bus or light rail: making the right choice' (Brighton: Environmental and Transport Planning, 2000).

Haughton, G., 'Developing sustainable urban development models', *Cities*, 13/4 (1997): 189–95.

Hawken, P., Lovins, A.B. and Lovins, L.H., *Natural Capitalism: the Next Industrial Revolution* (London: Earthscan, 1999).

Heil, M. and Pargal, S., 'Reducing air pollution from urban passenger transport: a framework for policy analysis', *World Bank Development Research Group Working Paper WPS1991* (Washington, DC, World, August 1998). Online (<http://www.worldbank.org/html/dec/Publications/Workpapers/WPS1900series/wps1991/wps1991.pdf> July 2001).

Her, K., 'Sustainable subways', *Taipei Review* (magazine), 51/7, July 2001. Online (<http://publish.gio.gov.tw/FCR/current/R0107p36.html>, July 2001).

Higashi, M., Sugaya, M., Veki, K. and Fujii, K., 'Plasma processing of exhaust gas from a diesel engine vehicle', *Proceedings of the International Conference on Plasma Chemistry*, 2, (1985): 366–71.

Higashi, M., Uchida, S., Suzuki, N. and Fujii, K., *Transactions of the Institute of Electrical Engineers Japan*, 111A, (1991): 457–73.

Hillman, M., Adams, J. and Whitelegg, J., *One False Move: a Study of Children's Independent Mobility* (London: Policy Studies Institute, 1990).

Hirst, P. and Thompson, G., *Globalization in Question: the International Economy and the Possibilities of Governance* (Cambridge: Polity Press, 1996).

Holten-Andersen, J., *Natur og Miljø: Påvirkninger og tilstand 1997* [*Nature and Environment: Impact and Conditions*] (Copenhagen: Ministry of Energy and the Environment together with Danmarks Miljøunder-søgelser, 1998).

Honda models [online]. Available from: http://www.honda2001.com/models/index.html (June 2001).

Hook, W., 'Role of non-motorised transportation and public transport in Japan's Economic Success', *Transportation Record*, 1441, (TRB, Washington, DC, 1994).
Hook, W. and Replogle, M., 'Motorisation and non-motorised transport in Asia', *Land Use Policy*, Vol. 13, No. 1, (1996): 69–84.
Hori, M., *Gendai Oshu no Kotsu Seisaku to Tetsudo Kaikaku: Joge Bunri to Opun Akusesu* (*Transport Policy and Railway Reform in Modern Europe: Vertical Disintegration and Open Access*) (Tokyo: Zeimu Keiri Kyokai, 2000).
Hu, G. and Kenworthy J., 'Threat to global survival? – a case study of land use and transportation patterns in Chinese cities', course material for City Policy online Masters Program at Murdoch University (1999).
Hu, G., *'Land Use and Transport in Chinese Cities – an International Comparative Study'*, unpublished University of Melbourne PhD thesis (2002 available from the Librarian, the Faculty of Architecture, Building and Planning, the University of Melbourne, Victoria, 3010, Australia).
Hubbert M.K., *Nuclear Energy and the Fossil Fuels*, American Petroleum Institute Drilling and Production Practice, Proceedings of the Spring Meeting, San Antonio, Texas, USA (1956): 7–25.
Hwang, Jim, 'The long way round', *Taipei Review* (magazine), 51 (7), July 2001. Online (<http://publish.gio.gov.tw/FCR/current/R0107p04.html> July 2001).
IP/01/997, Press release 'Clean urban transport: Euro 50 million for pilot cities', Brussels 16 July 2001.
IP/00/1198, 'Revitalising urban transport: EUR 50 million for pioneering cities', www.eurocities.org
IP/01/1263, *A Transport Policy for Europe's Citizens* (Brussels, 12 September 2001).
International Energy Agency, *Electric Vehicles: Technology, Performance and Potential* (Paris, Organisation for Economic Co-operation and Development Publication Service, 1993).
International Energy Agency, *International Energy Annual* (Washington, DC: IEA Energy Information Administration, 1999)
International Road Federation (IRF), *World Road Statistics 1990–1994* (Geneva and Washington, DC, International Road Federation, 1995).
IPCC (Intergovernmental Panel on Climate Change) *Third Assessment Report of Working Group 1* (<http://www.ipcc.ch/>, 2001).
Jackson, K.T., *Crabgrass frontier: the suburbanization of the United States* (New York: Oxford University Press, 1985).
Jacobs, M., *The Green Economy, Environment, Sustainable Development and the Politics of the Future* (London: Pluto Press, 1991).
Jamison, A., Eyerman, R. and Cramer, J. with Laessöe, J., *The Making of the New Environmental Consciousness: a Comparative Study of the Environmental Movements in Sweden, Denmark and the Netherlands* (Edinburgh: Edinburgh University Press, 1990).
Jänicke, M. and Weidner, H. (eds), *Successful Environmental Policy: a Critical Evaluation of 24 Cases* (Berlin: Sigma, 1995).
Japan Electric Vehicle Association (JEVA) (<http://jeva.or.jp> October 2000).
Kakumoto, R., *Kokutetsu Kaikaku: JR 10 Nen me Kara no Kensho* (*The Reform of JNR: A Retrospect from the 10th Anniversary of JR*) (Tokyo: Kotsu Shimbun Sha, 1996).
Kamioka, N., *Jidosha in Ikura Kakatte Iruka* ('How Much Does the Automobile Cost Us?' (Tokyo: Komonzu, 2002).
Kasai, T., *Mikan no Kokutetsu Kaikaku: Kyodai Soshiki no Hokai to Saisei* (*The JNR Reform Unfinished: from Collapse to Resurrection of a Colossal Organisation*) (Tokyo: Toyo Keizai Shimpo Sha, 2001).
Kato, H. and Sando, Y., *Kokutetsu, Denden, Sembai Saisei No Kozu* (*The Frames of Restructuring of the Railway, the Telegraph and Telephone and the Monopoly Public Corporations*) (Tokyo: Toyo Keizai Shimpo Sha, 1973).

Keen, S., *Debunking Economics: the Naked Emperor of the Social Sciences* (Annandale, Australia: Pluto Press, 2001, also published by Zed Books, London and St Martins Press, New York)

Kenworthy, J.R. and Laube, F.B., *An International Sourcebook of Automobile Dependence in Cities*, 1960–1990 (Boulder, Col.: University Press of Colorado, 1999).

Kenworthy, J.R. and Laube, F.B., *UITP Millennium Cities Database for Sustainable Transport* (CD-Database) (Brussels, International Union [Association] of Public Transport [UITP], 2001).

Kenworthy, J. and Newman P., *Automobile Dependence – the Irresistible Force?* (ISTP publication, Murdoch University, Perth, Australia, 1993).

Ko, S.F., *Shokuminchi Tetsudo to Minshu Seikatsu: Chosen, Taiwan, Chugoku Tohoku* (*The Colonial Railways and the Lives of the Grass-Roots: Korea, Taiwan and Manchuria*) (Tokyo: Hosei Daigaku Shuppankai, 1999).

Krugman, P. (1999) *The Accidental Theorist* (London: Penguin Books).

Kuhn, T.S., *The Structure of Scientific Revolutions* (Chicago and London: University of Chicago Press, 1962).

Laherrère, J.H., *Distributions De Type 'Fractal Parabolique' Dans La Nature* (Paris: Comptes Rendues, Académie de Science II 1996, 322).

Laherrère, J.H., 'World oil supply, what goes up must come down – but when will it peak ?', *Oil and Gas Journal*, 1 February (1999): 57–64.

Laird, P., Newman, P., Bachels, M. and Kenworthy, J., *Back on Track: Rethinking Transport Policy in Australia and New Zealand* (Sydney: UNSW Press, 2001).

Latour, B., *Science in Action: How to Follow Scientists and Engineers Through Society* (Cambridge, Mass.: Harvard University Press, 1987).

Le Clercq, F. (1987), 'Policy cycles and new planning Issues', in P. Nijkamp and S. Reichman (eds), *Transportation Planning in a Changing World* (Gower: Aldershot, 1987) pp. 93–108.

Lemmers, L., *'Amsterdam's traffic restraint policy'* (Amsterdam: Proceedings of the Car Free City Conference, *1994)*.

Lindblom, C.E., *Politics and Markets: the World's Political–Economic Systems* (New York: Basic Books, 1977).

Linn, J.F., *Cities in Developing Countries: Policies for Their Equitable and Efficient Growth* (Oxford: World Bank Research Publication, Oxford University Press, 1983).

Linstone, H.A. and Mistroff, J., *The Challenge of the 21st Century: Managing Technology and Ourselves in a Shrinking World* (New York: State University of New York Press, 1994).

Low, N.P., 'Ecosocialisation and environmental planning: a Polanyian approach', *Environment and Planning*, A 34/1 (2002).

Low, N.P. and Banerjee-Guha, S., 'Melbourne and Mumbai: contradictions of urban transport sustainability', paper presented at the '*World Planning Schools Congress on Planning for Cities in the 21st Century*', Tong-Ji University, Shanghai (2001).

Low, N.P and Gleeson, B.J., *Justice, Society and Nature: an Exploration of Political Ecology* (London and New York: Routledge, 1998).

Low, N.P. and Gleeson, B.J., 'Ecosocialisation or countermodernisation?: reviewing the shifting 'storylines' of transport planning', *International Journal of Urban and regional Planning*, 25/4 (2001): 784–803.

Low, N.P., Gleeson, B.J., Elander, I. and Lidskog, R. (eds), *Consuming Cities: the Urban Environment in the Global Economy After Rio* (London: Routledge, 2000).

Lundqvist, L.J., 'Sweden', in M. Jänicke and H. Weidner (eds), *National Environmental Policies: a Comparative Study of Capacity Building* (Berlin: Springer, 1997) pp. 45–71.

Luxembourg Nr 09403/98 (Presse 206, 17-06-1998).
McGowan, F., 'Transport policy', in D. Dinan (ed.), *Encyclopedia of the European Union* (London: Macmillan – now Palgrave Macmillan, 2000) pp. 460–2.
McKenzie, R.D. 'The ecological approach to the study of the human community' in Park, R.E., Burgess, E.W. and McKenzie, R.D., *The City* (Chicago: University of Chicago Press, 1967) pp. 63–79.
Maddison, D. *et al.*, *The True Costs of Road Transport*, Blueprint 5 (London: Earthscan, 1996).
Maezono, I., Chang, J.-S., 'Reduction of CO_2 from combustion gases by DC corona torches', *Institute of Electrical and Electronics Engineers, Transactions on Industrial Applications*, 26 (1990): 651–5.
Mahadevia, D., 'Sustainable urban development in India: an inclusive perspective', paper presented at the National Seminar on Globalisation and Environment (Mumbai: SNDT University, 2001): 1–32.
Mahmoudi, S., 'Environmental integration into transport policy – options for implementation' (Stockholm: Swedish EPA, 2000).
March, J.G. and Olsen, J.P., *Rediscovering Institutions: the Organisational Basis of Politics* (New York: The Free Press, 1989).
Martinez-Alier, J., *Ecological Economics, Energy, Environment and Society* (Oxford: Blackwell, 1987).
Marx, K., *Grundrisse: Foundations of the Critique of Political Economy* (Harmondsworth: Penguin Books, 1973).
Mees, P., 'A tale of two cities: urban transport, pollution and equality', in D. Glover and G. Patmore (eds), *New Voices for Social Democracy* (Sydney: Pluto Press for Australian Fabian Society, 1999) pp. 141–55.
Mees, P., *A Very Public Solution: Transport in the Dispersed City* (Melbourne, Australia: Melbourne University Press, 2000).
Metcalfe, S., 'Evolution and economic change', in A. Silburton (ed.), *Technology and Economic Progress* (London: Macmillan – now Palgrave Macmillan, 1990).
Metropolitan Regional Planning Authority, *The Corridor Plan for Perth* (Perth: Metropolitan Regional Planning Authority, 1970).
Metropolitan Town Planning Commission, *Plans of General Development: Melbourne* (Melbourne, Australia: Government Printer, 1929).
Miller, C., 'Australia worst on greenhouse', *The Age* (Melbourne) (November 1999): 3.
Ministerie van Verkeer en Waterstaat, *Beleidseffectmeting Verkeer en Vervoer*, annual reports [Ministry of Transport and Public Works, *Monitoring traffic and transport*] (The Hague, 1993 and 1995).
Mitsuzuka, H., *Kokutetsu wo Saiken Suru Hoho wa Kore Shika Nai* (*This is The Only Way to Restructure JNR*) (Tokyo: Seiji Koho Senta, 1974).
Mogridge, M.J.H., 'The self-defeating nature of urban road capacity policy: a review of theories, disputes and available evidence', *Transport Policy*, 4 (1997): 5–24.
Morris, L., 'Taking a toll', *Sydney Morning Herald* (15 January 2000): 41.
Moscovici, S., 'Des répresentations collectives aux répresentations sociales: éléments pour une histoire' [From collective representations to social representations: concepts for a historical account], in D. Jodelet (ed.), *Les Répresentations Sociales* [Social Representations] (Paris: Presses Universitaires de France, 1989): 62–86.
Motavalli, J., *Forward Drive* (San Francisco, CA: Sierra Club Books, 2000).
Mullally, G., 'Discourses on mobility at a European Level', in O'Mahony, Patrick (ed.), *ACT WILL Study 5 Evaluation of Possible Technological Options to Relive the Challenges Caused by the Saturation of Cities*. Cork: Centre for European Social Research, 1997).

MVA Consultancy, *Ho Chi Minh City Transport Study: Diagnostic Report* (Ho Chi Minh City, Transport and Urban Public Works Service of Ho Chi Minh City, Socialist Republic of Vietnam, 1997).

Nakanishi, K., *Sengo Nihon Kokuyu Tetsudo Ron* (*The Japan National Railways in the Post-War Period*) (Tokyo: Toyo Keizai Shimpo Sha, 1985).

National Audit Office (NAO), *Tackling Obesity in England* (London, National Audit Office, 2001).

National Resource Defense Council, *Aviation and the Environment* (Washington, DC, USA, National Resource Defense Council, 1996).

Nespor, J., *Knowledge in Motion: Space, Time and Curriculum in Undergraduate Physics and Management* (London and Washington: The Falmer Press, 1994).

Netherlands, Ministry of Transport and Public Works and the Ministry of Housing and Physical Planning, *Structuurschema Verkeer en Vervoer, deel a: beleidsvoornemen [Second Transport Structure Plan: Part a: a Draft of a Plan of Action]* (The Hague: Ministry of Transport and Public Works and the Ministry of Housing and Physical Planning, 1977).

Netherlands, Ministry of Transport and Public Works, *Tweede Struktuurschema Verkeer en Vervoer: deel a Beleidsvoornemen [Second Transport Structure Plan: Part a: a Draft of a Plan of Action]* (The Hague: Ministry of Transport and Public Works, 1988).

Netherlands, Ministry of Transport and Public Works and the Ministry of Housing, Physical Planning and the Environment, *Second Transport Structure Plan, Part D: Government Decision* (The Hague: Ministry of Transport and Public Works and the Ministry of Housing, Physical Planning and the Environment, 1990).

Network News, the Quarterly Magazine of the National Cycle Network, Issue No 6, 4th Quarter 1997.

Newman P., 'The rebirth of Perth's suburban railways', in D. Hedgcock and O.Yiftachel (eds.), *Urban and Regional Planning in Western Australia: Historical and Critical Perspectives* (Curtin University, Western Australia: Paradigm Press, 1992).

Newman, P.W.G. and Kenworthy, J.R., *Cities and Automobile Dependence: an International Sourcebook* (Aldershot: Gower, 1989).

Newman, P.W.G. and Kenworthy, J., *Sustainability and Cities: Overcoming Automobile Dependence* (Washington, DC: Island Press, 1999).

Newman, P.W.G., Kenworthy, J.R. and Laube, F., 'The global city and sustainability' (Jakarta, Indonesia: Fifth International Workshop on Technological Change and Urban Form, June 1997).

Noland, R., *Analysis of Metropolitan Highway Capacity and the Growth of Vehicle Miles of Travel*, Presentation to the Transportation Research Board, (January 2000).

Norquist, J.O., *The Wealth of Cities, Revitalizing the Centres of American Life* (Reading, Mass.: Addison Wesley, 1998).

NYMTC (New York Metropolitan Transportation Council) *Mobility for the Millennium: a Transportation Plan for the New York Region: Executive Summary* (New York: New York Metropolitan Transportation Council, 2000).

O'Connor, K., Stimson, R. and Daly, M., *Australia's Changing Economic Geography: a Society Dividing* (Melbourne: Oxford University Press, 2001).

OECD, *Market and Government Failures in Environment Management: the Case of Transport* (Paris: OECD, 1992).

OECD, *Urban Travel and Sustainable Development* (Paris: OECD, 1995)

OECD, *Pollution Prevention and Control: Environmental Criteria for Sustainable Transport*, Report on phase 1 of the project Environmentally Sustainable Transport (EST) (96) 136 (Paris: OECD, 1996).

Odell P.R., 'World oil resources, reserves and production', *The Energy Journal*, 15 (1994): 89–113.

Office of Transportation Technologies, *The FORD Program Accelerates Development of Hybrid Configurations* (1999a) (<http://www.ott.doe.gov/hev/> accessed June 2001)

Office of Transportation Technologies, *The General Motors/HEV Is Targeted for Consumer Acceptance* (1999b) (<http://www.ott.doe.gov/hev/> accessed June 2001).

O'Meara, M., 'Exploring a new vision for cities', in Brown, Flavin *et al.* (eds), *State of the World 1999*, (London: Earthscan, 1999) pp. 133–50.

O'Riordan, T., 'Democracy and the sustainability transition', in W.M. Lafferty and J. Meadowcroft (eds), *Democracy and the Environment: Problems and Prospects* (Cheltenham: Edgar Elgar, 1996) pp. 140–56.

ORTECH International, *Hybrid Electric Vehicles Final Report* (Ontario: Natural Resources Canada, 1993).

Palotai, T. and Chang, J.-S., 'Reduction of CO_2 from Ar–CO mixture gas by a Capillary Tube Reactor with Repeated Spark Discharges', *Proceedings of the 2nd International Symposium on High Pressure Low Temperature Plasma Chemistry* (1989): 143–50.

Patankar, P.G., 'Policy reforms for road development', paper presented at a seminar on 'Privatisation of Roads' (Mumbai: Indian Merchants Chamber, 1994).

Patankar, P.G., 'Urban transport: radical reorientation required', in *The Hindu Survey of the Environment 2000* (Chennai, 2000): 109–14.

Pearson, C.S., *Economics and the Global Environment* (New York: Cambridge University Press, 2000).

Perth Regional Transport Study Steering Committee, *Perth Regional Transport Study 1970* (Perth: Imperial Printing Company, 1971).

Peters, D., 'A sustainable transport convention for the new Europe' in F. Dodds, *Earth Summit 2002: a New Deal* (London: Earthscan, 2000) pp. 109–23.

Pharoah, T. and Apel, D., *Transport Concepts in European Cities* (Aldershot: Avebury, 1995).

Polanyi, K., *The Great Transformation: the Political and Economic Origins of our Time* (Boston, Beacon Press, 1957 [1944]).

Polèse, M. and Stren, R. (eds), *The Social Sustainability of Cities, Diversity and the Management of Change* (Toronto, University of Toronto Press, 2000).

Poth R., *Proceedings: CarFree City Conference* (Amsterdam, 1994): 45.

Pott, A., Doerk, T., Uhlenbusch, J., Ehlbeck, J., Hoschele, J. and Steinwandel, J., 'Polarization-sensitive coherent anti-Stokes Raman scattering applied to the detection of NO in a microwave discharge for reduction of NO', *Journal of Physics D: Applied Physics*, 31 (1998): 2485–98.

Prange, S., 'Both sides now – hybrid vehicles', *Home Power*, No. 82, (2001): 94–103.

Putnam, R., *Making Democracy Work: Civic Tradition in Modern Italy* (Princeton, NJ: Princeton University Press, 1993).

Rainbow, R. and Tan, H., 'Meeting the demand for mobility' (London: Selected Papers, Shell International, 1993).

Rand, D.A.J., Woods, R. and Dell, R.M., *Batteries for Electric Vehicles* (New York: John Wiley, 1998).

Rank, J., Folke, J. and Jespersen, P., 'Differences in cyclists and car driver exposure to air pollution from traffic in the city of Copenhagen', *The Science of the Total Environment*, 279 (2001): 131–6.

Reddy, C.M. 'Vehicle emissions', in *The Hindu Survey of the Environment 2000* (Chennai, 2000): 115–25.

Rifkin, J. *The Age of Access* (Tarcher/Putnam: 2000).

Rimmer, P.J., *Rikisha to Rapid Transport Urban Public Transport Systems and Policy in Southeast Asia* (Sydney: Pergamon Press, 1986).

Roberts, J., 'Changed travel ... better world? a study of travel patterns in Milton Keynes and Almere' (TEST Report No. 105 London: TEST 1991).

Roseland, M., *Toward Sustainable Communities: Resources for Citizens and their Governments* (Gabriola, BC: New Society Publishers, 1998).

Ries, I. 'City link: investor heaven, taxpayer hell', *Australian Financial Review* (15 December, 1995: back page and p. 29).

Rothengatter, W., 'Obstacles to the use of economic instruments in transport policy' in ECMT (European Conference of Ministers of Transport) *Internalisaing the Social Costs of Transport* (Paris: ECMT/OECD, 1994) pp. 113–52.

RTD info. Magazine for European Research., No. 32. December, 2001.

Rundell, G., 'Melbourne anti-freeway protests', *Urban Policy and Research* 3/4 (1985): 11–40.

Sachs, W., *Planet Dialectics: Explorations in Environment and Development* (London: Zed Books, 1999).

SACTRA (Standing Advisory Committee on Trunk Road Assessment), *Trunk Roads and the Generation of Traffic* (London: HMSO, 1994).

Saraf, R., 'Third world traffic', *The Hindu Survey of Environment, 1998* (Chennai, 1998) pp. 153–61.

Satterthwaite, D., 'For better living', *Down to Earth*, 5/5 (1996): 31–5.

Schafer, A. and Victor, D., *Scientific American* (special edition on transportation), (1997).

Scheurer J., (wwwistp.murdoch.edu.au 2001).

Searle, G., 'Sydney's recent transport planning: under economic ascendancy', paper presented to 15th Pacific Regional Science Conference Organization Wellington, New Zealand, December, 1997.

Searle, G., 'New roads, new rails lines, new profits: privatisation and Sydney's recent transport development', 17(2) (1999): 111–21.

Selman, P., *Local Sustainability: Managing and Planning Ecologically Sound Places* (London: Paul Chapman Publishing, 1996).

Sels, J.W. van Loben, *Interregional Transportation Strategic Plan* (California Department of Transportation, 1998). http:www.dot.ca.gov/ ... es/OASP/Interregional/Strategic.PDF. accessed February 2001).

Shanghai Star (newspaper), 'Shanghai eases its car market protectionism', 28 January 2000. Online (<http://www.chinadaily.com.cn/star/history/00-01-28/c13-car.html> July 2001).

Sharma, K., 'Vehicle pollution: deadman driving', *Hindu Survey of the Environment, 2000* (Chennai, 2000) pp. 105–8.

Sharma, R.S., 'Status of metropolitan transport in India' (New Delhi: The Times Research Foundation, 1985) pp. 1–88.

Sheller, M. and Urry, J., 'The city and the car', *International Journal of Urban and Regional Research*, 24/4 (2000): 737–57.

Simon, H., 'The fundamentals of e-business-strategy – part I', *Frankfurter Allgemeine Zeitung* (12 January 2002): 63.

Sims, W.A., 'Distributional issues in transportation pricing schemes for infrastructure and externalities', paper for technical seminar on 'Infrastructure pricing, investment decisions and financing' (Ottawa, Canada Royal Commission on National Transportation 1992), Vol. 20.

Sinclair, Knight, *Merz Environment Effects Statement, Scoresby Transport Corridor, Volume 1* (Melbourne: VicRoads, 1998).

Small, K.A., *Urban Transportation Economics* (Philadelphia: Harwood Academic Publishers, 1992).

Smith, M., Whitelegg, J. and Williams, N., *Greening the Built Environment* (London: Earthscan, 1999).

Soja, E.W., *Postmodern Geographies: the Reassertion of Space in Critical Theory* (London and New York: Verso, 1989).
Sone, S., 'Some ideas to enhance convenience of railways in co-operation with other modes of transportation', *Unyu to Keizai*, 62/3 (2002).
SOU, 'Ny kurs i trafikpolitiken' (1997): 35.
Spearitt, P. and deMarco, C., *Planning Sydney's Future* (Sydney: Allen and Unwin, 1988).
Sriraman, S., 'Road and road transport development in India', *Asian Transport Journal* (1997): 7–28.
Starkie, D., 'Configuring change: reflections on transport policy processes', in P. Nijkamp and S. Reichman (eds), *Transportation Planning in a Changing World* (Aldershot: Gower, 1987) pp. 269–83.
State Transport Study Group (STSG), *Sydney Urban Expansion Studies: Long Term Issue Study*, Report no. 218 (Sydney: Ministry of Transport, 1985).
Statistisk Årbog [*Statistical Yearbook*] (Copenhagen: Danmarks Statistik, 1995).
STPP (Surface Transportation Policy Project), *Easing the Burden: a Companion Analysis of the Texas Transportation Institute's Congestion Study* (2001).
Sweden, Ministry of Transport, *Sammanställning av Remissyttranden Över Kommunikationskommitténs Betänkande 'Ny Kurs i Trafikpolitiken'* [*A Survey of Rescripts from Various Authorities and Organizations on the Report of the Investigative Commission on the Future of Communication entitled 'New Directions for Transport Policy'*] (Stockholm, Sweden: Ministry of Transport, 1997).
Sweden, Sveriges Offentliga Utredningar, *Ny Kurs i Trafikpolitiken: Slutbetänkande* [*New Directions in Transport Policy: Final Report*] (Stockholm: Sveriges Offentliga Utredningar, 1997) p. 35.
Sweden, The Swedish National Road Administration, *Transportprognos 2005 och 2020* [*Transport Forecasts 2005 and 2020*] (Borlänge: The Swedish National Road Administration, undated).
Sweden, Transportpolitik för en hållbar utveckling [*Transport Policy for Sustainable Development*] The Swedish Government's Bill (Stockholm, 1997/98) p. 56.
Swedish National Road Administration, *Environmental Report* (Borlänge: Swedish National Road Administration, 1996).
SwEPA report no 4979, *Key Role-Players in the Process Towards Sustainable Transport in Europe: a Report from the Swedish Euro-EST Project*, Å.Vagland, and M, Viehauser (Stockholm: INREGIA AB, 1999).
SwEPA report no 5023, *Air Pollution from the Transport Sector – a Scenario Study for Europe*, G. Haq and P. Bailey (Stockholm: Stockholm Environment Institute, 1999).
SwEPA report no 5082, *Transport Infrastructure Financing on European Level* (Stockholm: Stockholm Environment Institute 2000).
SwEPA report no 5083, *Integrating Environment in Transport Policies – a Survey in EU Member States* (Stockholm: Stockholm Environment Institute, 2000).
SwEPA report no 5084, *EU – Fuel and Vehicle Tax Policy* (Stockholm: Stockholm Environment Institute, 2000).
T 2000 *Just the Ticket Traffic Reduction Through Parking Restraint* (London: Transport 2000, 1997a). Transport 2000 is at 12–18 Hoxton Road London N16NG.
T 2000 *Roads 21: a Roads Policy for the Next Century* (London: Transport 2000, 1997b). Transport 2000 is at 12–18 Hoxton Road London N16NG.
Tanaboriboon, Y., 'An overview and future direction of transport demand management in Asian metropolises', *Regional Development Dialogue*, 13/3 (1992): 46–70.
Tanner, R.J., 'Melbourne City Link: a financial analysis of operation after construction', 1996 mimeo in the possession of N. Low, Faculty of Architecture, Building and Planning, The University of Melbourne, 3010, Australia.

Tata Consultancy Services (TCS), *Strategic Options for Public Transport Improvement in Large Cities of Developing Countries*, Report for United Nations Centre for Human Settlements (1993): 1–96.

Tata Consultancy Services (TCS), 'Traffic, economic and environmental impact assessment of flyovers in Mumbai' (2001).

Taylor, K.C., *Proceeding of the First International Symposium Catalysis and Automotive Pollution Control (CAPOC 1)*, Brussels, 8–11 September 1986 (The Netherlands: Elsevier Science Publishers B.V, 1987).

Tengström, E. in collaboration with E. Gajewska and M. Thynell (1995), *Sustainable Mobility in Europe and the Role of the Automobile: a Critical Inquiry*, 2nd edn (Stockholm: The Swedish Transport and Communications Research Board, 1995) Report 17.

Tengström, E., *Towards Environmental Sustainability?: a Comparative Study of Danish, Dutch and Swedish Transport Policies in a European Context* (Aldershot: Ashgate, 1999).

TERM (European Environment Agency), *Indicators Tracking Transport and Environment Integration in the European Union* (Copenhagen: European Environment Agency, 2001).

TEST (1989), *Quality Streets – How Traditional Urban Centres Benefit from Traffic Calming* (London: TEST).

Teufel, D. *et al.*, *Oeko-Bilanzen von Fahrzeugen*, UPI Bericht No. 25, 6 (Heidelberg, Germany: Auflage, Umwelt und Prognose Institut, 1999).

Thomas, S., Zalbowitz, M., fuel cells – green power, Los Alamos National Laboratory, http://www.usfcc.com/Transportation.html [accessed 8 June 2001]. Toyota Prius [online] (<http://prius.toyota.com> June 2001).

Thomson, J.M., *Great Cities and Their Traffic* (London, Victor Gollancz, 1977).

Tiwari, G., 'Heterogenous cities', Hindu Survey of the Environment 1998 (Chennai, 1998): 141–5.

Town and Country Planning Board, Victoria, *Organisation for Strategic Planning: a Report to the Minister for Local Government in Response to his Letter of 3 May 1966* (Melbourne: Town and Country Planning Board, 1967).

Transport Foundation, *Fix Australia Fix the Roads*, Newsheet (January 2000).

Transurban, *The Melbourne City Link Prospectus* (Melbourne: Transurban City Link Ltd, 1996).

Tunali, O., 'Quest is on to beat traffic logjam', *The West Australian* (18 March 1996): 4/5.

Udall, R. and Andrews, S., 'Methane madness: a natural gas primer', *Solar Today*, 15/4 (2001): 36–9.

United States Department of Transportation (2000), 'The transportation equity act, for the twenty first century: users' guide on line, http://www.istea.org/guide/guideonline.htm (accessed September 2001).

Urashina, K., Miyamoto, H. and Ito, T., 'The reduction of NO_x from the diesel engine exhaust gas by superimposing barrier discharges', *Combustion Science and Technology*, 133, (1998): 79–91.

US Fuel Cell Council, 'Fuel cell power for vehicles', [online] http://www.usfcc.com (accessed June 2001).

USA, *Transportation Energy Data Book, Edn. 17* (Oak Ridge: National Laboratory for the United States Department of Energy, 1997).

USA Emission Standards, 'Cars and light-duty trucks' (Washington, DC: US Environmental Protection Authority, 1999).

USA, Central Intelligence Agency, *2001 World Fact Book* (Central Intelligence Agency, 2001).

Venkateswaralu, K., 'All for a whiff of fresh air', *Hindu Survey of Environment, Chennai 2000* (Chennai, 2000): 138–42.

Vickerman, R., 'Transport provision and regional development in Europe, towards a framework for appraisal' in Banister, D. (ed.), *Transport Policy and the Environment* (London and New York: E.F. and N. Spon, 1998) pp. 131–60.

Victoria, Infrastructure Planning Council, *Victoria 2020, Interim Report* (accessed October 2001: <http:www.dpc.vic.gov.au/ipc>.

Victorian Government, *Shaping Melbourne's Future* (Melbourne: Ministry for Planning and Environment, 1987).

Vuchic, V., *Transportation for Livable Cities* (New Brunswick, NJ: Center for Urban Policy Research, 1999).

Wainwright, R., 'Tunnel visions', *Sydney Morning Herald* (9 February, 2000a): 12.

Wainwright, R., 'Missing traffic hits toll revenues', *Sydney Morning Herald* (14 April, 2000b): 12.

Wang, G.T., 'Effective path – prior development of public transport, *City Planning Review*, 1 (1995): 7–10.

Wang, L.H. and Yeh, A. G.-O. (eds), *Keep a City Moving: Urban Transport Management in Hong Kong* (Tokyo, Asian Productivity Organization, 1993).

Webber, M.M., 'The post-city age', *Daedalus*, 97/4 (1968): 1093–9.

Weiss, H.R., 'Plasma induced dissociation of carbon dioxide', *Proceedings of the International Conference on Plasma Chemistry*, 2 (1985): 383–8.

White, R.R., 'Sustainable development in urban areas: an overview' in D. Devuyst *et al.* (eds), *How Green Is the City* (New York: Columbia University Press, 2000) pp. 47–62.

White, R.R., *Building the Ecological City* (Cambridge: Woodhead Publishing, 2002).

Whitelegg, J., *Driven to Destruction: Absurd Freight Movement and European Road Building* (Lancaster: Eco-Logica Ltd, 1994).

Whitelegg, J., *Critical Mass: Transport, Environment and Society in the Twenty First Century* (London and Chicago: Pluto Press, 1997).

Whitelegg, J. and Williams, N., *The Plane Truth: Aviation and the Environment* (London: Ashden Trust, 2000).

Willoughby, K., 'The "local milieux" of knowledge based industries', in Brotchie, J., Newton, P., Hall, P., Blakeley, E. and Battie, M. (eds), *Cities in Competition* (Melbourne, Australia: Cheshire, 1994).

Winger, A.R., 'Finally: a withering away of Cities?', *Futures*, 29/3 (1997): 251–6.

World Bank, *Sustainable Transport: Priorities for Policy Reform* (Washington DC: World Bank, 1996).

World Health Organization (WHO), *Community Noise*, Environmental Health Criteria Document (Copenhagen, Denmark: World Health Organization, 1993).

Yada, T., 21 *Seiki no Kokudo Kozo to Kokudo Seisaku (The Structure of National Land and National Land Policy in 21st Century)* (Tokyo: Taimeido, 1999).

Yasoshima, Y. (ed.), *Sogo Kotsu Repoto 2: Zenkoku Ichinichi Kotsuken* (*A Comprehensive Transport Report 2: Covering the Entire Nation within the Range of a Day Trip*) (Tokyo: Gyosei, 1988).

Yee, C.Y., *Crash Course in Bay Area Transportation Investment* (San Francisco: Urban Habitat Program, 1999).

Yiftachel, O. and Kenworthy, J., 'The planning of metropolitan perth: some critical observations', in D. Hedgcock and O. Yiftachel (eds), *Urban and Regional Planning in Western Australia: Historical and Critical Perspectives* (Curtin University, Western Australia: Paradigm Press, 1992).

Youngquist, W., *Geodestinies: the Inevitable Control of Earth Resources Over Nations and Individuals* (Portland, OR: National Book Company, 1997).

Index

References to figures, tables and boxes are in *italics*; those for notes are followed by 'n'

CPI Antony Rowe
Chippenham, UK
2017-07-07 10:28